HYDRAULIC FLUIDS

A GUIDE TO SELECTION, TEST METHODS, AND USE

M. Radhakrishnan

ASME PRESS
NEW YORK
2003

Copyright © 2003
The American Society of Mechanical Engineers
Three Park Ave., New York, NY 10016

Library of Congress Cataloging-in-Publication Data

Radhakrishnan, M. (Mariappa), 1940–
    Hydraulic fluids: a guide to selection, test methods, and use/M. Radhakrishnan.
    p. cm.
    ISBN 0-7918-0184-5
    1. Hydraulic fluids.   2. Fluid power technology.   I. Title.

TJ844.R33 2003
621.2′042 – dc21

2002038203

To

my parents and
sister, Pappakka

CONTENTS

Chapter 6 Test Methods on Fire Resistant Hydraulic Fluids 91

Chapter 7 Biodegradable Hydraulic Fluids 111

Chapter 8 Maintenance of Hydraulic Fluids 137

Chapter 9 Disposal, Regulations, Reclamation, and Rerefining 163

Epilogue   181

Appendices   183

References   231
The Author   253
Index   255

ACKNOWLEDGMENTS

The ASTM Standards materials used in this book have been adapted with the permission of ASTM. The complete ASTM Standards may be purchased from ASTM, 100 Barr Harbor Drive, West Conshohocken, PA 19428; phone: 610-832-9585; fax: 610-832-9555; e-mail: *service@astm.org*; web site: www.astm.org.

The DIN Standards extracts used in this book have been reproduced with the permission of *DIN Deutsche Institut fur Nornung e.V.* The definitive version for the implementation of these standard is the edition bearing the most recent data of issue, obtainable from Beuth Verlag GmbH, Burggrafenstrasse 6, 10787 Berlin, Germany.

The terms and definitions taken from ISO 3448:1992, ISO 4404:1998, ISO 4406:1999, ISO 6072:1999, ISO 6743-4:1999, ISO 13357-2:1998, ISO 14935:1998 are reproduced with the permission of the International Organization for Standardization (ISO). These standards can be obtained from any ISO member and ISO Central Secretariat web site at *http://www.iso.ch*

Extracts from the 7th Fire Resistant Hydraulic Fluids Report of Luxembourg Test Procedures are reproduced with the permission of the European Commission, Employment and Social Affairs DG, Luxembourg.

The excerpt of the CEC-L-33-A-93 Test Procedure is reproduced with the permission of the Coordinating European Council for the Development of Performance Tests for Transportation Fuels, Lubricants and Other Fluids, U.K.

The IP standards extracts used in this book have been reproduced with the permission of The Institute of Petroleum, 61, New Cavendish Street, London W1G 7AR, U.K.

Thanks are also due to Denison Hydraulics Inc., U.S.A.; Cincinnati Machine, A UNOVA Company, U.S.A.; Eaton Corporation, U.S.A.; ExxonMobil Lubricants & Specialties, U.S.A.; Factory Mutual Research, U.S.A.; the Organization for Economic Cooperation and Development (OECD), France; the Society of Automotive Engineers, U.S.A.; and the VDMA, Germany, for permission to use excerpts from their standards/publications.

The following organizations were kind enough to provide me the materials relating to hydraulic fluids, which helped in many ways in the preparation of this book.

Ashok Leyland Limited, Hosur, India
BP Oil International (U.K.) Limited

The British Fluid Power Association, U.K.
Central Power Research Institute, Bangalore, India
Cincinnati Machine, A UNOVA Company, U.S.A.
Denison Hydraulics Inc., U.S.A. and Denison Hydraulics France S.A.
Eaton Corporation, U.S.A
ExxonMobil Lubricants & Specialties, U.S.A. and U.K.
FMC Corporation (U.K.) Limited
Factory Mutual Research, U.S.A.
Fenner Fluid Power Limited, U.K.
Fuchs Petrolube A.G., Germany
Great Lakes Chemical Corporation, U.K.
Hausinco Water Hydraulics Limited, U.K.
Indian Institute of Petroleum, Dehra Dun, India
Leonardo Engineering Pvt.Limited, Bangalore, India
Mannesmann Rexroth A.G., Germany
Oiltech Australia Pty. Limited Australia
Organization for Economic Cooperation and Development (OECD),
    France
Pall Industrial Hydraulics, U.K.
Rhein Chemie Rheinau GmbH, Germany
Stanhope-Seta Limited, U.K.
Viscolube Italiana, Italy
Yuken India Limited, Bangalore, India

When I started this work, I was a little fearful about whether I could fulfill the high standards demanded of it. But the enthusiasm and support of my friends Dr. N. Raman, Mr. G. Jayaraman, and Mr. V.R. Sritharan were a source of great strength to me. Dr. Raman and Mr. Sridharan also took pains to go through the manuscript and gave me valuable suggestions to shape the book into a better form. I also needed access to good technical library facilities, which was generously provided by the Central Manufacturing Technology Institute (CMTI), Bangalore; the Indian Institute of Science, Bangalore, and the National Aerospace Laboratories, Bangalore: my sincere thanks to all of them. I am very much indebted to Mr. Y. Kokubun, who shared all his experiences with me and who was kind enough to permit me to use his case studies in the book. My ex-colleagues Mr. S. G. Venkata Prasad, Mr. B. Chandrasekhar, Mr. B. R. Mohan Raj, and Mr. P. S. Ranganatha Rao did all the illustrations with the fervor of love and regards–my special thanks to them. For want of some key papers, I went through many anxious moments. I thank very much Mr. David Philips, Mr. Renato Schieppati, Mr. Sudhir Singhal, Dr. Seichi Wattanabe, and Mr. V.V. Pattanashetti for relieving me those anxious moments. Thanks are also extended to Mr. Gerry Saunders, Dr. H. N. Chanakya, Mr. S. S. Radhakrishnan, Mr. S. Varadhachari, Dr. E. Radha, and Mr. T. R. Prasanna, for providing technical support. I thank all the reviewers for their valuable comments, which made the content of the book richer. Thanks are also due to Professor Duncan Dowson and Ms. Sheril Leich for helping to bring the manuscript to print form and also to ASME International, and in particular to my editor Ms. Mary Grace Stefanchik at ASME Press, for the interest taken in the publication of this book. Thanks are also due to Mr. Colin McAteer and his colleagues for making the book more readable by excellent copy-editing and production.

The extensive time I spent on the computer perplexed my wife, Kalaiselvi, and my family members. Their forbearance and understanding, made this project possible. Lastly, I would like to acknowledge that it was Mr. M.E. Visveswaran and the late Mr.R.K.Gejji ex-directors of CMTI, who were instrumental in giving me the tribologist avatar, which I cherish very much.

PREFACE

Among industrial lubricants, the consumption of hydraulic fluids is highest at both the industry level and the national level. Besides being used in hydraulic systems, they are also widely used for the lubrication of rolling bearings, fluid film bearings, and, to some extent, even gears. Hence a better understanding and proper care of hydraulic fluids are very important for performance as well as for economic considerations. But there is an apparent lack of appreciation of the effective use of hydraulic fluids. During my association with many training programs relating to hydraulics and lubrication, I have observed that the participants always exhibited immense enthusiasm to know and learn about hydraulic fluids or lubricants. It is an encouraging sign that will put spirit into every tribologist. It will augur well if that positive spirit could be transformed into definite results in terms of higher performance or better productivity of equipment. Over time, I recognized that the absence of a comprehensive reference on hydraulic fluids was a serious handicap to getting the best performance out of hydraulic fluids. It was this reflection that prompted me to begin consolidating all the aspects of hydraulic fluids and work toward shaping it as a book. When the Fluid Power Society of India (FPSI) at Bangalore also came to me with a similar proposal, it gave me added incentive. This book is the outcome of that.

The book covers a broad spectrum of hydraulic fluids that are widely used in industrial hydraulics. For easy reference and detailed treatment, the subject is covered in nine chapters. After the introductory chapter, a detailed treatment on mineral oils, the most extensively used hydraulic fluid, is given in Chapter 2. As most of the properties covered are also applicable to other lubricants, readers may find it very useful. Practicing engineers, in general, are hardly familiar with the various test procedures relating to hydraulic fluids. A sound background knowledge of hydraulic fluid test procedures can greatly help users in the selection, application, and maintenance of hydraulic fluids. Hence a conscious effort has been made to include all the relevant tests relating to hydraulic fluids, and they are covered in Chapter 3. High water content fluid (HWCF) has been given exclusive coverage in Chapter 4, without grouping it with fire-resistant fluids. As many users may still consider using HWCF as a replacement of mineral oils rather than as a fire resistant fluid, readers may find the separate treatment on HWCF informative.

Hydraulically operated devices now replace operations that were performed manually near a heat or fire source, and hence the use of fire-resistant fluids is expanding. Tighter state regulations and better understanding of fire hazards, and the consequent development of test procedures, have made significant contributions toward the development of fire-resistant hydraulic fluids. Keeping these developments in mind, fire-resistant fluids and the various test procedures associated with them are covered in Chapters 5 and 6, respectively.

An emerging technology in hydraulic fluids is biodegradable fluids. In recent years, environmental protection has become a major global issue. This increased awareness has given impetus to the development of biodegradable hydraulic fluids. Major lubricant manufacturers are quite responsive to today's social and engineering needs. As a result, commercially, a range of biodegradable fluids is now available. To apprise readers of this new development, a concise account of biodegradable fluids has been given in Chapter 7, which I hope will help readers understand the nuances of this new technology.

Maintenance, an important component of any industrial system, often is not appreciated in practice. In the case of hydraulic fluids, it helps not only to extend the longevity of the fluid but also the associated equipment. Hence a comprehensive treatment on maintenance of all hydraulic fluids is given in Chapter 8. Improper disposal of used oils and its impact on the environment are drawing the attention of many countries, and disposal of used oils is being regulated through state legislation. From both the economical and environmental perspective, reclamation and rerefining of hydraulic oil have assumed special importance. In order to emphasize these important socioeconomic aspects, the last chapter is fully devoted to disposal, regulations, and recycling of used oils. Data sheets on commercially available antiwear mineral oils, fire-resistant fluids, and biodegradable fluids serve as a handy reference to readers for the selection as well as the application of hydraulic fluids. They will also be helpful in understanding the specifications and the various properties of hydraulic fluids. Readers may find certain details and data repetitive; but their inclusion is done to make each write-up as self-contained as far as possible and to free readers from frequent jumping to pages for reference.

For a variety of reasons most practicing engineers tend to think that hydraulic fluids or lubricants are shrouded in mystery. This may be due to the fact that mechanical engineers, who are the prime users of hydraulic fluids and lubricants, tend to view it as not their cup of tea and to look upon it as an alien subject—a discipline of chemical engineering. As a result, past practice and manufacturers' recommendations continue to reign over the lubrication practice in industry. Keeping this in mind, the chemistry of fluids is dealt with to the minimum extent possible.

The success of any engineering solution lies in how well factors such as the operating and environmental conditions, economic considerations, organizational culture, and prevailing maintenance practice are integrated into the technical solution. The data and the guidelines given in this book, I trust, will help readers diagnose and find solutions to the problems concerning hydraulic fluids in particular and lubricants in general.

The book caters to the needs of beginning as well as experienced engineers. Also, I hope that those who did not have the opportunity to pursue formal engineering education will find the book helpful while dealing with hydraulic fluids and lubricants in their work. A book cannot fulfill all the needs of all readers. However, I expect that readers will find an answer to most of the questions that arise in their minds. If any readers do not find an answer to their questions or the answers given are not adequate, I request they kindly bring these instances to my notice so that future editions can fill the gap.

M. Radhakrishnan
Bangalore, India
December, 2002

Chapter 1

INTRODUCTION

Background

Blaise Pascal discovered the power of fluids around the year 1650. He propounded a theory that the pressure developed on a confined fluid acts equally through the fluid in all directions. The mechanical lever was the only known device till then for moving heavy masses in order to gain mechanical advantage. Pascal's discovery gave birth to the "fluid lever," or, in today's parlance, "fluid power," for force amplification (fig.1–1).

As with many discoveries, Pascal's was not put into use for a long time. The first known application of Pascal's law was the hydraulically operated Bramah Press, which used water as the hydraulic fluid for power transmission—100 years after Pascal's time [1]. The application of fluid power since then has expanded leaps and bounds, in practically all conceivable engineering devices—from hi-tech aerospace applications to entertainment artifacts, machine tools to mobile equipment, and sugar mills to simple loading platforms.

The functions, operating conditions, and the types and ranges of fluid power machinery are so varied now that the fluids used in hydraulic systems over the years has also gone through many development stages to meet such varied performance requirements. Consequently, many types of hydraulic fluids have also emerged. Hydraulic fluid is a vital component of the hydraulic system, and it will be seen that no other component in the hydraulic system is as multifunctional as the fluid.

Functions of Hydraulic Fluids

The primary function of hydraulic fluid is to transmit power from the pump to various actuating elements, such as hydraulic motors and cylinders, through pressure and flow. To meet this performance criterion, the fluid used should possess low compressibility and foam tendency and good flow characteristics over the operating temperature range. There are some secondary functions, however, which are essential for the smooth and successful operation of the hydraulic system. They are:

As a seal between moving parts: The fluid should have the required viscosity to provide adequate sealing, thereby minimizing leakage across the internal

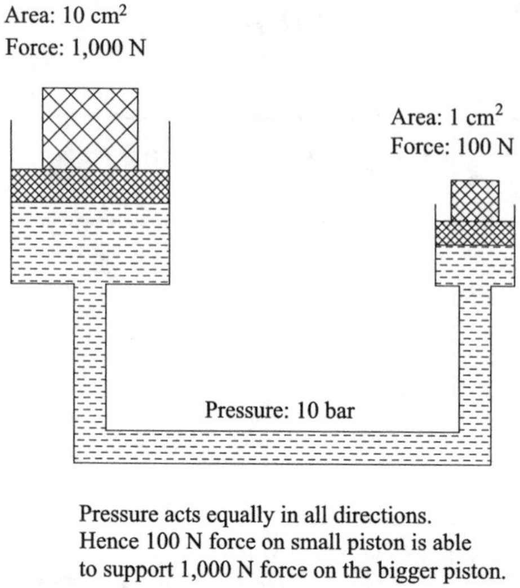

Area: 10 cm^2
Force: 1,000 N

Area: 1 cm^2
Force: 100 N

Pressure: 10 bar

Pressure acts equally in all directions.
Hence 100 N force on small piston is able
to support 1,000 N force on the bigger piston.

FIGURE 1–1 Pascal's Principle

pressure gradient in the pumps and valves. Adequate viscosity and shear stability are the properties that are required to satisfy this function.

As a lubricant: This is perhaps the most important function of the fluid. To reduce friction and wear of moving parts in the pumps, motors, and valves, the fluid film interposed between the sliding elements should have adequate strength and lubricity. Many properties of hydraulic fluids, such as viscosity, antiwear, thermal and hydrolytic stability, demulsibility, filterability, and oxidation resistance are developed to satisfy this function.

As a heat transfer medium: As the fluid passes from the pump to various actuators and valves, heat is generated within the system. This heat must be effectively transported so that system performance remains consistent and reliable, as the temperature rise in the fluid can significantly alter many performance factors. It is believed that in any hydraulic system, 15 to 20 percent of the power is converted into heat. Properties such as specific heat and thermal conductivity of the fluid therefore must be high enough to make the heat transfer effective.

As a corrosion and rust inhibitor: The fluid should be capable of protecting the surface of the various metals used in the hydraulic system as well as in the machine from corrosion-inducing elements such as water, heat, and oxygen. Also, it should be compatible with the various metals used in the hydraulics and machine. In other words, the chemistry of the fluid should not induce any corrosion on the normally used metals.

Some hydraulic fluids may also need special properties such as fire and radiation resistance. In addition, in recent times there is a new emphasis on properties such as low toxicity and ecological acceptability. It may look rather strange to a user that current development work on hydraulic fluids is largely centered on improving performance relating to the secondary functions of the oil. This is probably because the primary functions of the fluid are largely determined by the inherent property of the base fluid used.

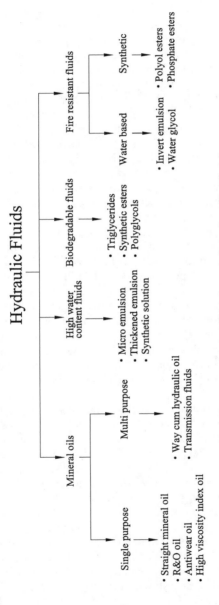

FIGURE 1-2 Family of Industrial Hydraulic Fluids

Types of Hydraulic Fluids

Water was the earliest fluid used in hydraulic systems. Despite its low viscosity, poor lubricating property, corrosiveness, and limited working temperature range, it was used until the 1920s, when mineral oil appeared as an alternative fluid. The use of water as a hydraulic medium then slowly receded and was restricted to systems that are either large or required fire resistance. Recent trends, however, indicate that water as a hydraulic fluid is getting a new lease on life as its use expands into meat processing, textile and wood industries, fire fighting rescue tools, robots used for abattoirs and radioactive areas, and washing and mixer units [2–8]. It is of interest to note that currently available water hydraulics hardwares claim to be suitable for an operating pressure of as high as 140 bar and even for servo applications.

Mineral oils used in the 1920s and 1930s contained no additives and are often referred to as straight mineral oils. In the 1940s the demand for high-performance fluids led to the appearance of rust and oxidation (R& O) inhibited oil. In the 1950s fire resistant fluid was developed, as the fire resisting property of mineral oil was found to be inadequate. Both synthetic and water-based fluids came to the market. Then, in the 1960s, to meet higher wear performance requirements, antiwear (AW) hydraulic oil was introduced, which is the most widely used fluid today. In the mid-1970s high water content fluid (HWCF) was introduced in the market to counter the high cost of mineral oil brought about by the oil crisis. The development work on HWCF was mainly spearheaded by the U.S. Now HWCF is used not only where fire risks exist but also where economic considerations are important. Many other multipurpose hydraulic oils, such as slide way cum hydraulic oil, and common fluid for transmissions and hydraulics, were also introduced.

Synthetic hydraulic fluids came into the market in the mid-1930s to address severe operating conditions, such as wide operating temperature range, better thermal stability, and oxidation resistance coupled with fire resistance requirements. They are more expensive than petroleum-based oils and are generally used in mobile and aerospace applications. In industrial applications, use of synthetic fluids is limited to areas where the fire resistant property is a prime requirement. Growing global concerns about environmental protection have triggered the development of biodegradable hydraulic fluids. Many biodegradable fluids are now commercially available [9]. These fluids are in the introduction stage and largely confined to applications where they are strictly required. However, in Europe they are used even for the lubrication of gears and metal cutting operations; their use is likely to expand in years to come.

As hydraulic equipment was updated in terms of power and performance from the 1940s on, the quality and performance level of mineral oils were also upgraded continuously to match changing performance needs. As said earlier, applications of fluid power are so widespread and varied, and the performance requirements of hydraulic fluids have become so diverse, that the types of hydraulic fluids available are also on the increase to meet specific performance needs. The family of industrial hydraulic fluids given in fig.1–2 illustrates this point, the details of which will be discussed in succeeding chapters.

Chapter 2

MINERAL HYDRAULIC OILS

Hydraulic fluids are essentially nothing but base fluids blended with a host of chemicals, called additives, to impart or enhance desired properties and performance characteristics. Base fluids can be either from petroleum stock or synthetic fluids. Mineral hydraulic oils are derivatives from petroleum stock and are basically hydrocarbons—compounds of carbon and hydrogen. Over 95 percent of the lubricants in use today are made from petroleum stock; among all the hydraulic fluids, mineral oil is the most widely used. The reasons are:

- It is available in a wide range of viscosity, and hence users have a wide choice of selection for any application.
- Its performance can be enhanced or modified by blending with suitable additives.
- It is chemically inert and compatible with a variety of materials and other industrial lubricants.
- It can be used for lubrication of other machine elements, for example bearings and gears, thus helping in rationalization of lubricants.
- It requires a less rigorous maintenance schedule and retains its properties over a long period of time.
- It is cost-effective.

REFINING OF PETROLEUM CRUDE

Petroleum stock or crude oil is obtained by drilling into the earth's crust as deep as six to eight kilometers. The crude oil excavated from the earth is refined to form the base oil, which is used for blending various types of lubricants, including mineral hydraulic oils. Crude oil is a complex mixture of many different compounds, with a predominance of hydrocarbons—varying from simple paraffin hydrocarbons to the more complex aromatics. Besides water, salt, and other siliceous materials, many compounds of sulphur, nitrogen, oxygen, and metals such as nickel and vanadium are also present in crude oil. The refining process recovers the various products contained in the crude.

The technology of refining is complex because the products recovered are quite varied and large, ranging from light hydrocarbon gases to heavy asphalts and coke, and many proprietary processes are in existence [1]. The desired end product also influences the type of refining process used. The constituents of the crude oil obtained from various sources can differ very much, and the refining process is modified to suit such contingencies so as to maintain consistency in the quality of the base oil produced. Understandably, given the wide range of products that come out of refining, it is beyond the scope of this book to delve into all the details of refining. The process outlined here therefore covers only the typical steps that go into the production of hydraulic oil and lubricants.

The initial steps in any refining process of crude oil are desalting and dewatering operations. The crude oil is heated, and an emulsion breaker is added. Settling and filtering remove the water and salt contained in the crude from the oil phase. The other important steps that follow dewatering and desalting are:

- **Atmospheric distillation** Crude oils contain hydrocarbons with a wide range of boiling points, and the first step is to separate substances with a different boiling range. Separation is done by atmospheric distillation. After the desalting operation, the crude oil is heated under pressure (little more than the atmospheric pressure) to a temperature of around 340 °C (644 °F). The mixture of hot liquid and vapor enter the fractionating column, where hydrocarbons of different boiling range are separated. The different yields are collected in the atmospheric tower.
- **Vacuum distillation** The residue obtained from the atmospheric distillation of the crude (where groups of hydrocarbons are separated from their boiling range) is fed into the vacuum distillation column for further processing. As different hydrocarbons boil at different temperatures, distillation can effectively separate them into different fractions. Lubricating oils are thermally sensitive and decompose easily at temperatures above 340°C (644°F), and hence vacuum distillation (about 15 to 50 millibar) is done, as a low temperature is only required at low pressures. Here a series of fractions representing different molecular ranges or viscosity ranges are separated out through the distillation column, and flash point and viscosity are adjusted for characterization of the fractions.
- **Solvent extraction** In this step, in simple terms it can be said that aromatic compounds are separated from nonaromatic compounds. The intensity of the process depends upon the power of the solvent, temperature, and solvent-to-oil ratio used. Different yields can be extracted, and it is sufficient for us to know that at this stage aromatics and paraffinic hydrocarbons are separated out. Thermal and oxidation stability, and viscosity-temperature characteristics of the extracts are improved in this process.
- **Dewaxing** Wax present in long-chain normal paraffins can lead to a high pour point or poor flow characteristics at low temperatures. However, these long-chain paraffins are essential to good viscosity-temperature characteristics. Hence the base oils are only partially

dewaxed, and the desired pour point is achieved by pour point depressants. One way to remove this wax is to mix the oil with methyl ethyl keytone and then chill the mixture to a temperature of around $-12°C$ to $-6°C$ ($10°F$ to $21°F$). The wax crystals are precipitated and separated out from the oil by filtration.

- **Finishing process** There are many finishing processes, but the most widely used is known as hydrogenation, which has practically replaced the old clay treatment. Many proprietary hydrofinishing processes exist. Hydrogenation is not new—it dates back to the 1920s—but because of high cost it was not favored [4] for long. Several types of hydrogenation process are available depending on the nature of feedstock and the end use of the products. Better yield can be realized in hydrogenation, as it is a conversion process, unlike clay treatment, which is a separation process. Hydrogenation basically involves passing the heated oil and hydrogen through a catalyst bed to remove odor, color, nitrogen, and sulfur. A recent process is hydrocracking—a severe hydrogenation process [2,3], which involves a mixed phase of hydrogen and gaseous and liquid hydrocarbons flowing down through multiple fixed beds of catalysts at a pressure of around 200 bar (2900 psi) and a temperature over $350°C$ ($662°F$). In hydrocracking, aromatics can be converted into naphthenes, and paraffinic molecules can be rearranged to get a lube stock with better temperature-viscosity characteristics, and improved thermal and oxidation stability. An added advantage is that the lube stock produced by hydrocracking responds better to additives. This process can produce lube stock of superior quality, as it is less dependent on the quality of the crude.
- **Blending** Blending these refined base oils with suitable additives makes the finished lubricants or hydraulic oils. Additives provide the required performance characteristics. Viscosity is also adjusted at this stage by mixing different base oils. Blending is done normally at a temperature of $50°C$ to $60°C$ ($122°F$ to $140°F$) to ensure satisfactory mixing without causing any thermal stress to the oil.

As said earlier, petroleum oils are complex mixtures containing many hydrocarbons; mixtures of paraffins, naphthenes, aromatics, and the typical molecular structures, are given in fig. 2–1. Lubricating oils may typically contain 20 to 40 carbon atoms in each molecule. Physical properties and the performance characteristics of hydraulic oils depend on the relative distribution of paraffinic, aromatic, and naphthenic components. For example, n-paraffin exhibits a high melting point, low density, low viscosity, and little change of viscosity with temperature. In contrast, relatively lower melting points, greater viscosity, and a larger change in viscosity with temperature are the characteristics of isoparaffin. Parraffinic oils also possess better oxidation stability, but their high melting point causes them to crystallize as wax. On the other hand, pure aromatics possess higher viscosity but poor viscosity-temperature characteristics, and density is higher than that of the paraffinic and naphthenic

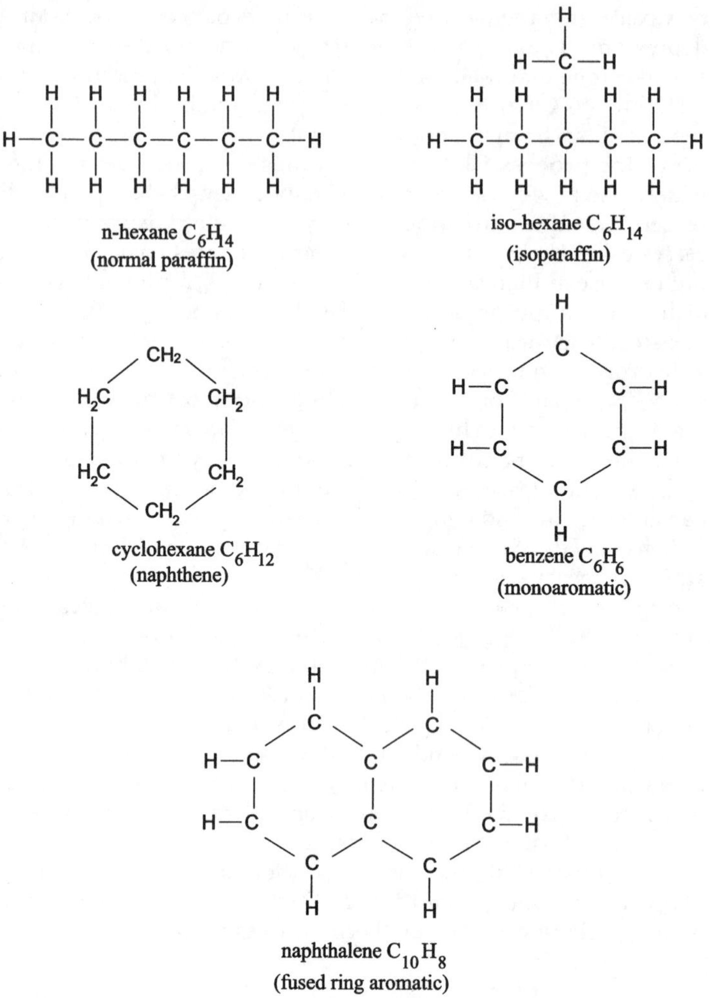

FIGURE 2–1 Different Configurations of Hydrocarbons

oils. A notable feature of aromatics is its high solvency power, which makes the blending of additives easier. Most of the characteristics of naphthenic oils are in between those of paraffins and aromatics. An advantage of naphthenic oils are their low melting point, which does not contribute to wax formation.

Because of marketing considerations, oils are produced in three types: High Viscosity Index (HVI) oils, Medium Viscosity Index (MVI) oils and Low Viscosity Index (LVI) oils. HVI oils are produced from crude oils containing more paraffins, while LVI oils contain more aromatics and MVI oils contain more naphthenes. The lubricating oil is classified as paraffinic, naphthenic, and aromatic, depending upon the type of structure that is predominant. Premium lubricants as well as hydraulic oils are made from paraffinic base oils (HVI oils) consisting primarily of paraffinic structures with naphthenes and aromatics in small proportions. Now let us look at the various types of mineral hydraulic oils. Broadly speaking there are two types: single-purpose and multipurpose mineral hydraulic oils.

SINGLE-PURPOSE MINERAL OIL

Straight Mineral Oil

Straight mineral oil or plain mineral oil, as the name implies, contains no additives. It was the first replacement fluid for water and was used until the 1930s. Plain mineral oil is available in a range of viscosity grades. Per the ISO (International Standards Organization) classification [5], these oils come under the type HH. Details of the ISO classification are given in Table A-4 in Appendix 4. As many high-performance fluids have emerged, the use of straight mineral oil is now limited to jacks and small presses only. It is used, however, as a lubricant in many applications, even, for instance, in the lubrication of gears.

Rust and Oxidation (R&O) Inhibited Oil

This type of oil, originally used in turbines, came to hydraulics and is fortified with rust and oxidation inhibitor additives. Per the ISO classification, these oils come under the type HL. They are also available in a range of viscosity grades. A comparative specification of R&O hydraulic oil is given in Table 2–1 [6–8]. R&O oils marketed today also possess antifoam and demulsibility characteristics. Where the loads and pressures are moderate, R&O oils adequately meet performance criteria. In hydraulics they are used in many piston pumps and gear pump applications, and, to a lesser extent, in vane pumps. Though they are not considered high-performance oils, they are quite widely used as a lubricant in many circulating systems. Hydraulics apart, they are also used in turbine and compressor applications.

Antiwear Oil

R&O oil blended with antiwear additives is known as antiwear type oil. Per the ISO classification, these oils come under the type HM. A comparative specification of antiwear hydraulic oil is given in Table 2–2 [6–8]. It is the most widely used hydraulic oil and was introduced in the 1960s to meet the then-demanding applications of high speed, pressure, and power that were encountered in hydraulic systems of the day. Its availability in a range of viscosity grades and its high performance level made it useful to many applications. It is also used for the lubrication of rolling as well as fluid film bearings, and also for the lubrication of spur and helical gears. Data sheets on commercially available antiwear mineral oils are given in Appendix 1 from Tables A1–1 to A1–16 [9].

Table 2–1 Comparative Specifications of R&O Hydraulic Oil

Sl. No.	Parameters	Denison HF-1	Cincinnati machine P-38/55/54	DIN 51524 PART 1 HL
1	Viscosity index, minimum	90	90	(See note 4)
2	Rust ASTM D 665 A/DIN 51585,	Pass	Pass	0-A
	ASTM D 665 B	Pass	–	
	Copper corrosion DIN 51759			100 A3
3	Oxidation stability,			
	(ASTM D 4310/DIN 51587)			
	After 1,000 hours, TAN increase	1.0	–	2.0
	Total insoluble sludge, mg	100	–	–
	Copper, mg	200	–	–
4	Hydrolytic stability ASTM D 2619			
	Copper weight loss, mg/cm^2	0.2	–	–
	Acidity of water layer (TAN)	4.0	–	–
5	Air release, minutes, DIN 51381	–	–	5–VG 10, 22, and 32
				10–VG 46, and 68
				14–VG 100
6	Foam test ASTM D 892/DIN 51566	None	–	Nil
	Foam volume after 10 minutes, mL			
7	Demulsibility, minutes	40-37-3	–	30–VG 10
	(ASTM D 1401/DIN 51599)			40–VG 22, 32, and 46
				60–VG 68, and 100
8	Thermal stability			
	168h @135°C, Cu + Fe rod			
	Sludge/100mL, mg	100	25	–
	Copper weight loss/200mL, mg	10	10	–
	Steel weight loss/200mL, mg	–	1	–
	Viscosity change %	–	5	–
	TAN increase	–	0.15	–
9	Axial pump wear test, Denison	Pass	–	–
	(Test being revised)			
10	Neutralization number (TAN)	–	0.2	Report
	ASTM D 974/DIN 51558			
11	Seal compatibility			VG 32/46 VG 68/100
	(DIN 53521, and DIN 53505)			
	Relative change in volume%	–	–	0 to 12 0 to 10
	Change in shore A hardness	–	–	0 to −7 0 to −6

Notes:
1. Figures given are maximum allowable value unless otherwise specified.
2. DIN standard numbers given are meant for DIN specification only.
3. Thermal stability test (Sl.No.8) is per Cincinnati Machine's Procedure A.
4. Instead of VI, DIN specifies the viscosity at −20°C, 0°C, 40°C, and 100°C.
5. Denison specifications are under revision.

High Viscosity Index (HVI) Oil

Though the base oil used for hydraulics by itself possesses a high viscosity index, certain applications demand a viscosity index that is even higher. Viscosity improvers are therefore added to the oil to enhance its viscosity index [10]. Basically they are HM type oils fortified with VI improvers and are classified as

ISO type HV. Now, per the revised ISO classification [7], R&O oil fortified with VI improvers, class HR, has also been introduced. As HR type is a recently introduced one, it has yet to make its presence felt in the market. Commonly used viscosity improvers are high molecular weight linear polymers, such as polymethacrylates or polyalkylstyrenes. Like HM and HL type oils, they are also available in a range of viscosity grades. Due to the presence of VI improvers, at low temperatures the oils exhibit a low viscosity to facilitate low friction; the same lubricant, however, maintains a significant viscosity at high temperatures so that adequate fluid film thickness can be maintained. Though VI of HVI oils is generally around 150, oils with VI as high as 200 are also available.

Most industrial hydraulics applications generally do not need hydraulic oil fortified with VI improvers. Use of these oils is limited to special applications only, for example, in equipment that is used outdoors and is subject to wide temperature changes, or in machine tools, or in the lubrication of fluid film bearings where change in viscosity has to be kept to a minimum under the entire range of operating temperatures in order to maintain the desired performance level. Since the viscosity change with respect to temperature is minimal with HVI oils, it can make pump start-up even at a low temperatures easier. For example, to give an idea of the change in viscosity level of HVI oils at two extreme temperatures, typically the viscosity of HVI oils grade VG 32 of VI 150 at 0°C (32°F) is around 240, compared with 350 of the same grade HM type oils. But the change in viscosity at 100°C (212°F) is not pronounced—the viscosity of HM type comes down to around 5.2, compared with 6.3 of HV type oil at 100°C, hardly a difference of 1.1 cSt.

MULTIPURPOSE MINERAL OIL

It is desirable to have one fluid serve many purposes, as it can help variety reduction and rationalization of lubricants. Fewer lubricants in a single machine, or in an industry as a whole, mean less chances of misapplication and easing lubrication management. Also, the problem of different oils getting mixed under the operating conditions due to human error need not be a serious concern for the users. Because of these concerns, multipurpose oils such as slide way cum hydraulic oil, transmission fluids, hydraulic cum air compressor oil, and hydraulic cum gear oil came into use. Among these, slide way cum hydraulic oil is the one most widely used in industry.

Slide Way Cum Hydraulic Oil

This is a combination fluid suitable for lubrication of machine tool slide ways as well as hydraulic systems and is classified as a type HG oil. Unlike other types of oils, this type is available only in ISO grades VG 32 and 68, and to some extent in VG 220. In other words, it is available only in the viscosity grades that are normally used in machine tool slide ways. In antiwear type oil, polar additives are added to enhance the boundary lubricating property of oil so that slide ways can be operated free of stick-slip; this means slide ways can be smoothly

Table 2-2 Comparative Specifications of Antiwear Hydraulic Oil

Sl. No.	Parameters	Denison HF-0	Denison A-HF2	Cincinnati machine P-68/69/70	DIN 51524 PART 2 HLP
1	Viscosity index, minimum	90	90	90	(See note 4)
2	Rust ASTM D 665 A/DIN 51585	Pass	Pass	Pass	0–A
	ASTM D 665 B	Pass	Pass	–	
3	Copper corrosion DIN 51759				100 A 3
	Foam test ASTM D 892/DIN 51566				
4	Foam volume after 10 minutes	None	None	Nil	Nil
	Demulsibility, minutes	40-37-3	40-37-3	–	30–VG 10
	(ASTM D 1401/DIN 51599)				40–VG 22, 32, and 46
					60–VG 68, and 100
5	Oxidation stability				
	(ASTM D-4310/DIN 53587)				
	After 1,000 hours, TAN increase	1.0	1.0	–	2.0
	Total insoluble sludge, mg	100	100	–	–
	Copper, mg	200	200	–	–
6	Hydrolytic stability ASTM D 2619				
	Copper weight loss, mg/cm^2	0.2	0.2	–	–
	Acidity of water layer (TAN)	4.0	4.0	–	–
7	Air release, minutes, DIN 51381	–	–	–	5–VG 10, 22, and 32
					10–VG 46, and 68
					14–VG 100
8	Thermal stability				
	168 hours @135°C, Cu + Fe rod				
	Sludge/100 mL, mg	100	100	25	–
	Copper weight loss/200 mL, mg	10	1.0	10	–
	Steel weight loss/200 mL, mg	–	–	1	–
	Viscosity change %	–	–	5	–
	TAN change %	–	–	±50	–

No.	Characteristics			
9	Vane pump wear test, ASTM D 2882 Total weight loss (ring and vane), mg	–	50	Ring-120 (DIN 51389) Vane-30
9A	Vane pump wear test, Denison T6C-020	Pass	–	–
9B	Axial pump wear test, Denison	Pass	–	–
	Load capacity FZG gear test DIN 51354 Part 2			(Test applicable for VG 32 to 100 only)
	Minimum damage power stage		–	10
10	Neutralization number (ASTM D 974/DIN 51558 Part 1)		1.5	Report
11	Filterability test Denison TP 02100 Filtration time, seconds			
	a) without water	600(x)	600(x)	–
	b) with 2% water	< 2x	< 2x	–
12	Seal compatibility DIN 53521 and 53505			VG 32/46 VG 68/100
	Relative change in volume %		–	0 to 12 0 to 10
	Change in shore A hardness		–	0 to –7 0 to –6

Notes:

1. Figures given are maximum allowable value unless otherwise specified.
2. Thermal stability test (Sl.No.8) is per Cincinnati Machine's Procedure A.
3. Instead of VI, DIN specifies the viscosity at −20°C, 0°C, 40°C, and 100°C.
4. DIN standard numbers given is meant for DIN specification only.
5. Denison specifications are under revision.

run at very low as well as high speeds, without any jerky motion. The polar additives normally used are hydrocarbons derived from natural products such as fatty oils, fatty alcohols, or esters of fatty acids (glycerides of rapeseed or lard oil). As the polar additives contain fats that are prone to easy oxidation, this type of oil has relatively shorter life than single-purpose hydraulic oil. Hence the bulk oil temperature should be limited to say, 50°C (122°F), compared with 60°C (140°F) of normal hydraulic oil, if the oil change interval has to be kept high. The advantage of using this oil is that it helps in reducing the variety of oils used in a machine tool. This oil is also suitable for lubricating lead screws. As lead screws and slide ways are always close to one another, it is convenient to combine the lubrication of both these elements with single oil.

When this oil is used for combined lubrication of slide ways and hydraulics, slide ways are also lubricated by the same circulating system. The oil, which returns from slide ways to the reservoir, is prone to contamination by metal cutting chips, cutting fluids, and other contaminants from external sources. The continuous ingress of these contaminants accelerates oil degradation, and frequent oil change is a common phenomenon in such systems. In the design stage itself, therefore, precaution must be taken to protect slide ways from contaminants; otherwise any gain from using single oil will be lost. If protecting slide ways from contaminants is a difficult proposition, it is advisable to separate the hydraulic system from slide way lubrication and lubricate the slide ways separately with a slide way oil using a total loss system. Bear in mind that any multipurpose lubricant generally balances the various lubrication requirements, and often there is a trade-off between various needs, which of course; in general, applies to any multipurpose product. Hence, when using multipurpose lubricants, the limitations have to be carefully taken into account.

Transmission Fluids

The types of hydraulic oils discussed so far are used where the fluid power energy is transmitted through pressurized fluid flow—these are hydrostatic applications. Transmission fluids, on the other hand, are used where the fluid velocity energy produces rotary motion or output torque—these are hydrokinetic applications. ISO classification for transmission fluids is HA for automatic transmissions, and HN for fluid couplers. The commonly known devices that use hydrokinetic energy are fluid couplings and torque converters. Compared with hydrostatic applications, hydrokinetic applications are few and far between. Transmission fluids are primarily used in manual and automatic transmissions in automotive vehicles and farm equipment. However, their use has spread to other industrial applications that were perhaps not thought of by formulators [17–23]; for instance, off-highway construction machinery, marine hydraulic systems, rotary screw type flood lubricated air compressors, rotary compressors, hydraulic hoists, hydraulic clutches in drilling rigs, mining equipment, and even gearboxes in drilling machines. In a way, it can be said that transmission fluids are used in applications where the oil is subject to arduous duty, such as high pressure, high speeds and low ambient temperatures.

In the classical sense, transmission fluids are truly multifunctional or multipurpose—used for both hydrokinetic as well as hydrostatic applications. Their multifunctional characteristics can be seen in a vehicle operating with an automatic transmission system: The fluid has to operate brakes and multiplate clutches, power steering, and torque converters; it must lubricate varieties of gears and bearings; and act as a medium for hydraulic control. To meet such a variety of requirements, the composition of the fluid has to be a complex one, and it is: it contains as many as 15 additives. Unlike normal hydraulic fluids, it has to withstand high speeds, loads, and temperatures, and also should possess high oxidation stability.

Viscosity for automatic transmission fluids (ATF) is quoted at 100°C, as the operating temperature of the fluid is much higher than that of other hydraulic systems; this viscosity generally corresponds to either ISO VG 32 or 46 grades. Some of the ATFs are also available in SAE grades. The viscosity index is generally more than 150. Another important feature of ATF is that it possesses excellent low temperature properties; the pour point is around −40°C (−40°F).

Besides the various properties that are required for a normal hydraulic fluid, the one that distinguishes transmission fluids from other hydraulic fluids or high-performance lubricants is the special requirement of friction characteristic [19–23]. The clutch operation in the transmission system has to be free of shudder, chatter, and squawk, which means that the torque converter output shaft has to be driven with minimal slip under fluctuating crankshaft speed conditions. To fulfill this function, the oil should primarily provide minimum variation in the value of coefficient of friction at all speeds in the clutch operation, which could arise due to variations in temperature, viscosity and contact pressure, and oxidation of the oil. Ideally, if the transmission is to be free from a "clunk" noise, the value of the static and dynamic coefficient of friction have to be almost the same. Differences in design from manufacturer to manufacturer, and the consequent different materials used in the clutch—for example, sintered material, graphite composites, paper composites, and asbestos and ceramic composites—compound the problem, i.e., the fluid formulation has to suit the particular material used. This performance characteristic is achieved by fortifying the oil with the so-called friction modifier additive, which has to perform its function under prolonged high shear and high temperature conditions. The additives used are oil-soluble polar compounds (fatty acids), such as sulfurized natural fatty acids, phosphoric acids, amines, and amides, and they have the capacity to form a thin chemisorbed film on the sliding surface, which can shear more easily during sliding, thus reducing friction. The type of additive and its concentration varies from one formulation to another, and this additive must also be compatible with others used in the oil. It is a complex technique; in fact, it is a technology by itself and requires extensive testing to assess performance characteristics.

The specifications and test methods for transmission fluids must conform to the standards set by the auto majors like GM, Ford, Chrysler, and Mercedes Benz. Dexron III (GM), Mercon (Ford), ATF + 3 (Chrysler), and C-4 (Allison Transmission) are some of the known specifications of the Original Equipment Manufacturer (OEM). In Europe and Japan, the fluid used must have the approval of OEMs. Most of the test methods used for normal hydraulic fluids (discussed in the next chapter) are not applicable for ATFs.

PROPERTIES OF HYDRAULIC FLUIDS

Because hydraulic fluids have many functions and must work under a wide variety of operating and environmental conditions, the number of properties they must possess also increased over the years. The principal required properties are dictated by the demands of the primary and secondary functions of hydraulic fluids, and are listed below:

- Viscosity / Viscosity index
- Density
- Bulk modulus
- Foaming resistance
- Air release property
- Oxidation stability
- Antiwear performance
- Filterability
- Pour point
- Rust and corrosion inhibition
- Compatibility
- Demulsibility
- Hydrolytic stability
- Thermal stability

Along with the above properties, special ones such as shear stability, fire resistance, biodegradability, radiation resistance, and dielectric properties are also required, depending upon the applications. Mineral oils do not have some of these properties. Fire resistance and biodegradability are dealt with in later chapters; the fourteen properties listed above will be discussed now. It is pertinent to note here that most of these properties should also be part of any high-performance lubricant. Typical values for important physical properties are given in Table 2–3.

VISCOSITY

In simple terms, the viscosity of an oil is the measure of its resistance to flow, which is caused by internal fluid friction. Low viscosity oil flows freely, and high viscosity oil flows sluggishly. The concept of viscosity is illustrated in fig. 2–2. It can be seen from the figure that the oil in contact with the moving surface moves with the same velocity (u) as that of the surface, and that the oil which is in contact with the stationery surface is at zero velocity. The force required to move the plate is proportional to the viscosity or internal friction of the oil. When the moving surface area is 1 cm^2, the film thickness is 1 cm, the velocity 1 cm/sec, and the required force dragging the surface is 1 dyne, then the viscosity

Table 2-3   Important Physical Properties of Mineral Hydraulic Oils

Sl. No.	Properties	Value
1	Base oil	Paraffinic
2	Specific gravity @ 15°C	0.86–0.89
3	Vapor pressure @ 50°C, bar	1×10^{-6}
4	Surface tension N/m	23×10^{-3}
5	Viscosity @ 40°C, cSt	22 to 100
6	Viscosity index, minimum	90
7	Flash point (COC) °C, minimum	200
8	Autoignition temperature °C	350
9	Pour point °C	−20
10	Thermal conductivity @ 20°C W/mk	0.134
11	Specific heat C_p@ 25°C J/kg °C	1.89
12	Bulk modulus @ 0-350 bar	
	@ 25°C in MPa	2,000
	@ 40°C in MPa	1,870
13	Coefficient of thermal expansion (volumetric)/ °C	0.0007

Note: Values given are only typical.

is said to be 1 poise. Since poise is a large unit, centipoise (cP) is the preferred unit. For example, water at 20°C has a viscosity of 1 cP. From this it follows that friction is related to viscosity, and the force required to move the plate is proportional to viscosity. This viscosity is called dynamic or absolute viscosity. The value of this viscosity is used in all design calculations, e.g., in bearing design and oil flow calculations.

Viscosity can also be explained by Newton's hypothesis. Newton discovered the relationship between shear stress and the rate at which fluid deforms. A solid subjected to shear will deform by a finite amount, whereas a fluid subjected to shear stress will deform indefinitely without limit, no matter how small the applied shear stress.

A fluid is subjected to shear stress when there is a velocity gradient normal to the direction of the motion. Consider an element of fluid in such a flow

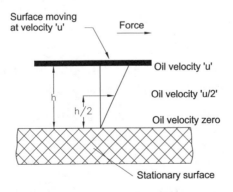

FIGURE 2-2   Concept of Viscosity

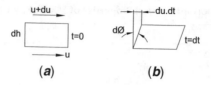

FIGURE 2–3a and 2–3b   Newton's Hypothesis

(fig. 2–3a and 2–3b) and apply the above relationship:

$\tau \; \alpha \; (d\varphi / dt)$ where τ = shear stress, ϕ = shear strain
from fig. 2–3a and 2–3b
$d\phi$ is small, hence $\tan d\phi = d\phi$ radians
$d\phi = du \, (dt / dh)$
$d\varphi / dt = du / dh$
$\tau \; \alpha \; (d\varphi / dt)$ or $\tau = \eta \, (du / dh)$

where η is the constant of proportionality and is called the coefficient of dynamic viscosity. In other words, viscosity is essentially independent of the rate of shear. The fluids that follow this relationship are called Newtonian fluids. Industrial mineral hydraulic oils are all Newtonian fluids; the unit of viscosity is poise.

$$1 \text{poise} = 10^{-1} \text{Ns} / \text{m}^2 \text{ or kg} / \text{ms (pascal seconds)}$$
$$1 \text{centipoise (cP)} = 0.01 \text{poise (milli pascal seconds)}$$

Kinematic Viscosity The instruments required to measure absolute viscosity are based on the flow that is produced by a constant pressure difference, and they are not simple. In addition, auxiliary equipment is needed. Because the flow that can be effected through a capillary under the action of gravity is more convenient, the viscosity measured under this condition is called kinematic viscosity. Kinematic viscometers are cheap and easy to operate, hence only kinematic viscosity is used in the specification of hydraulic fluids and other lubricants. This gravity flow is dependent on the density of the liquid. Kinematic viscosity (v) is given by the relation:

$v = \eta / \rho$, where ρ is the fluid density
The unit of kinematic viscosity is the stoke
1 stoke = 10^{-4} m^2 / second

Again, since the stoke is a large unit, kinematic viscosity is indicated by centistokes

$$1 \text{ centistoke (cSt)} = 10^{-6} \text{ m}^2 / \text{second}$$

In accordance with ISO recommendation [24], viscosities of all industrial lubricants are specified in centistokes at 40°C, as 40°C is reasonably close to the

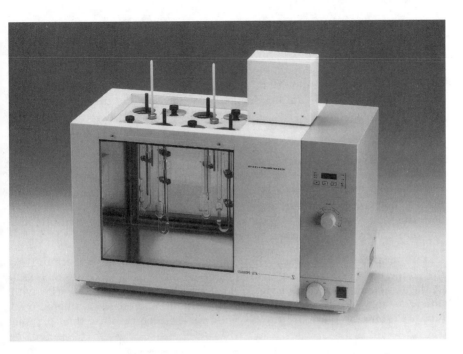

FIGURE 2–4 Constant Temperature Bath with Viscometers (Courtesy: Stanhope-Seta Limited, U.K.)

operating temperature of many applications. ISO viscosity grades range from 2 cSt to 1,500 cSt in 18 grades (Appendix 5, Table A5) in a geometric series, the viscosity of each successive grade being approximately 50 percent higher. A manufacturing tolerance of ± 10 percent of the nominal viscosity is allowed. The viscosities of all industrial hydraulic fluids are quoted per ISO standards only. Nevertheless, manufacturers do supply lubricants outside the ISO grades for special applications, but users may not get the same easy availability and economical price that are available with standard grades. It may be noted that because viscosity changes with temperature, viscosities are always quoted at a specific temperature, and therefore viscosity measurements are made in a constant temperature bath. The constant temperature bath apparatus shown in fig. 2–4 can maintain the temperature within ± 0.01°C.

In practice, viscosity is also expressed in Saybolt Universal Seconds (SUS) in the U.S., Redwood seconds (R1) in the U.K., and Engler degrees (°E) in continental Europe. They are empirical, and the conversion table is given in Appendix 6 (Table A6), where the equivalent kinematic viscosity in cSt for different systems can be found. Users are cautioned that when comparisons are made, they should be done at the same temperature.

A Note on SAE Viscosity Classification Many users are confused by SAE (Society of Automotive Engineers, U.S.) viscosity numbers. The SAE system is meant only for engine oils and automotive gear oils. SAE oils do not refer to a viscosity but to a narrow range of viscosity specified at 100°C (210°F)—the typical temperature of a hot engine. The viscosity of SAE 30 grade oil, for example, ranges between 9.3 and 12.5 cSt at 100°C. It is also obvious that the SAE system cannot be adopted for hydraulic oils, as the operating temperature (40°C/104°F) assumed in hydraulics is much less than that of the SAE system (100°C). Also, modern hydraulics demand oil with viscosities that have a narrow band of tolerance.

Factors Affecting Viscosity

Viscosity being the principal property of the hydraulic fluid governing the performance criteria of the hydraulic system, it is important to be thoroughly familiar with the factors that affect the value of viscosity. Temperature, pressure, and rate of shear are the three main factors that affect viscosity.

Temperature The viscosity of oil changes markedly with temperature. As the temperature increases, viscosity decreases; conversely, viscosity increases as the temperature decreases (fig. 2–5). This change in viscosity varies with different types of oils and with different refining methods. Naphthenic base oils, for example, thin more rapidly with a rise in temperature than paraffin

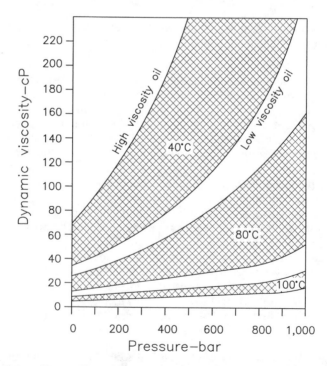

FIGURE 2–5 Effect of Temperature and Pressure on Viscosity (Courtesy: © BP Oil International Limited, U.K.)

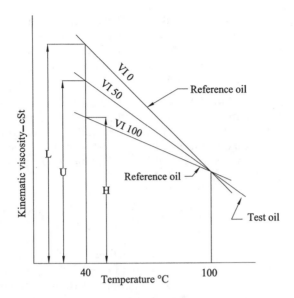

FIGURE 2–6   Calculation of Viscosity Index

base oils. This temperature effect on viscosity brings us to another property of the fluid known as viscosity index.

Viscosity Index   Viscosity index (VI) is widely used as a measure of change in viscosity with temperature. Viscosity index is an arbitrary scale based on two reference oils, one with a high rate of change (VI = 0) and the other with a low rate of change (VI = 100). Any oil quality that falls between these two oils (VI 0 and VI 100) gets a VI number between 0 and 100, depending upon its characteristics. The computation of VI can be explained with reference to fig. 2–6. The top line in the figure represents the oil with 0 viscosity index, and the lower line represents the oil with 100 viscosity index. The middle line between represents the oil whose viscosity index is unknown. The selected two reference oils and the unknown oil have the same viscosity at 100°C.

The VI of the unknown oil is given by the expression:

$$VI = \frac{L - U}{L - H} \times 100$$

Where L = Viscosity of the 0 VI oil at40°C in cSt
H = Viscosity of the 100 VI oil at 40°C in cSt
U = Viscosity of the unknown oil at 40°C in cSt

The value of L and H are selected from the tables given in any national standards such as ASTM D 2270 or BS 4459 against the value of the test oil's viscosity at 100°C. To find the VI of any oil, first find its viscosity at 40°C and 100°C in cSt. For example, if the viscosity of the unknown oil at 40°C and 100°C is 32 and 5.3 respectively, select the value of L and H against the value of 5.3 cSt from the tables; in this case it is 45.53 and 31.40. Then calculate the VI per the above formula, which works out to a VI of 95.75, say, 96. Since only high viscosity index (HVI) oils of paraffin base are used for hydraulics, the minimum VI of any hydraulic oil is 90. By virtue of the basis of calculation, readers may note that the VI of any oil cannot exceed 100.

Extended Viscosity Index Improvements in manufacturing techniques, the introduction of synthetic lubricants, and the use of VI improvers as one of the additives to enhance the value of VI brought about a situation in which the values of VI exceeded the traditional limit of 100. In order to meet present-day needs, a new method has been designed to calculate VI when it exceeds 100; it is known as extended viscosity index VI_E:

$$(K V_{100})^n = H / U$$

$$VI_E = \frac{(\text{antilog } n) - 1}{0.0075} + 100$$

$$\text{where } n = \frac{\log H - \log U}{\log K V_{100}}$$

$K V_{100}$ = kinematic viscosity of the sample oil at 100°C

H = kinematic viscosity at 40°C of an oil that has a VI of 100 and the same viscosity at 100°C as the sample

U = kinematic viscosity of the sample at 40°C

n = the power required to raise the sample viscosity at 100°C to equal the ratio of H / U

Shear Stability As mentioned earlier, mineral hydraulic oils are Newtonian fluids, but when high molecular weight polymers are added as VI improvers, the oil exhibits non-Newtonian character. What does that mean? When the oil is subjected to high rates of shear—for example, at the tip of the vane in the vane pumps or gears in gear pumps and throttling in flow control valves—the oil exhibits a drop in viscosity, but it recovers as soon as the low shear condition is restored. This phenomenon is known as temporary loss of viscosity. However, bear in mind that there are situations that can lead to a permanent loss of viscosity, and discerning users should take this factor into account when using oil with VI improvers. Nevertheless, it is claimed that VI improvers used in recent times have more shear stability [10], so as to reduce temporary viscosity loss at a high rate of shear due to alignment of molecules and permanent viscosity loss due to rupture of molecules. Assessing the shear stability property of hydraulic oils is dealt with in the next chapter.

Effect of Pressure on Viscosity It is known that lubricants and hydraulic oils exhibit a notable increase in viscosity when they are subjected to high pressure. Oil can become plastic solid, say, at pressures of around 1200 bar (17,400 psi). The normal hydraulic oil's viscosity is doubled when the pressure is raised to, say, 350 bar (5075 psi). However, at the lower end of pressures, say around 100 bar (many system pressures in industrial hydraulics are within this limit), the increase in viscosity (fig. 2–5) particularly with HVI oil is small and can be ignored. In other words, fluids that show a large change of viscosity with temperature also show a large change with pressure. But in high-pressure applications, such as extrusion machines, this effect can be more pronounced and may cause problems. In such applications, synthetic fluids that are less affected by pressure are used.

The increase in viscosity due to pressure affects three performance parameters: it increases the load-carrying capacity of the load-bearing elements such as bearings, gears, or slide ways; it increases friction losses in the system; and it increases the film thickness between sliding elements. Increase in load capacity and film thickness is beneficial, whereas increase in frictional loss is not. The designer should weigh the advantages and disadvantages carefully when designing high-pressure hydraulic systems, taking into account the effect of pressure on the viscosity of oils.

Influence of Viscosity on Performance Factors

As indicated earlier, viscosity is the most dominant property of the hydraulic oils and lubricating oils. A host of performance factors are dependent on viscosity, as illustrated in fig. 2–7. A detailed account follows.

Fluid Film Formation According to the definition of viscosity, because of internal friction, oil can be drawn into the clearance between the sliding elements; for example, between the spool and body or between the pump body and vane. The thin film thus formed can prevent metal-to-metal contact, reducing friction and wear. Oil with a high viscosity is more effective for this purpose than low viscosity oil, as high viscosity oil gives a thicker film.

Leakage As mentioned earlier, one of the functions of hydraulic oil is to act as a seal between sliding members, such as in spool type valves, pumps, and

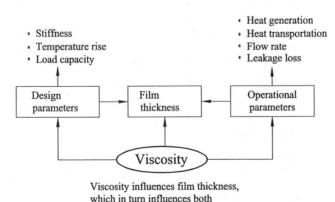

Viscosity influences film thickness, which in turn influences both design and operational parameters.

FIGURE 2–7 **Influence of Viscosity**

motors. The situation encountered here is similar to a close clearance seal—resistance to flow depends directly on oil viscosity. From this it follows that leakage and pressure loss will be greater with low viscosity. The leakage could be more significant in piston pumps, where leakage increases as the cube of clearance. Judicious selection of viscosity in such applications is important to improve the performance of the pump. Also, leakage in the hydraulic system can significantly lower performance; for example, it can cause sluggish movement or poor response in actuators like cylinders.

System Response High viscosity oil can cause unacceptable friction in actuating elements such as hydraulic cylinders, which can cause the actuation of hydraulic cylinders to become sluggish. For better response, low viscosity oils are preferred, particularly where servos are used. The viscosity of the oil normally used in servo application is 22 cSt or 32 cSt. Where high response with less leakage is required, high viscosity oil should be compensated with bigger valves and piping to get the desired response.

Pressure Loss When oil passes through valves and tubing, pressure loss occurs due to fluid friction and fluid velocity. This loss is proportional to the viscosity as well as to the velocity of the fluid. Again the selection of viscosity is the trade-off between leakage, pressure loss, and heat generation.

Heat Generation High fluid friction not only causes pressure loss and sluggish movement, it can also generate considerable heat, raising the oil temperature. It is pertinent to note that temperature rise is one of the undesirable elements of any hydraulic system. Besides reducing the life of the oil, in machine tools, for example, temperature rise can cause significant thermal deformation in the machine so as to affect the accuracy of the machined workpiece. Often temperature rise in the hydraulic oil becomes an important limiting factor in achieving the desired accuracy of the components in machine tools. Also, temperature rise in oils and the consequent fall in viscosity can cause inconsistency in the feed movement of machine tool slides, particularly when the slide is operating in the fine feed mode. Controlling heat generation from the hydraulics is therefore often critical to achieving higher performance in machine tools. The fall in viscosity also increases leakage significantly, particularly in high pressure systems such as presses, which can result in reduction of available power. Heat generation is greater with high viscosity oil than with low viscosity oil. Selecting oil with the right viscosity is important for efficient functioning of the hydraulic system.

SELECTION OF VISCOSITY

Viscosity is selected against many conflicting requirements such as leakage, film strength, pressure loss, heat generation, system response, and operating and ambient temperatures. Given the performance requirements and the various limitations imposed by the oil, the designer must know where to trade off. Apart from the factors given above, the hydraulic pump operating conditions also influence the selection of viscosity. Pump manufacturers recommend

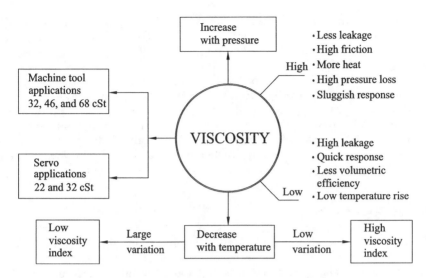

FIGURE 2–8 Factors Influencing the Selection of Viscosity

maximum and minimum viscosity ranges of the oil during start-up and during running for commonly used pumps such as gear, vane, and piston pumps. For example, the typical recommended maximum viscosity at start-up for inline piston pumps is around 220 cSt (1000 SUS); for vane and angle pistons, it is around 860 cSt (4000 SUS) [25]. This value can vary from one manufacturer to another. Over the entire operating temperature range, the selected viscosity should be within the limits of start-up and running viscosity range. The factors just discussed would help in the selection of the right viscosity for a given application. For most industrial applications, viscosity generally lies between VG 22 and VG 100 grades. In machine tool application, for example, the commonly used viscosities are VG 32, 46, and 68. Past practice and test results can also help enormously in choosing the right viscosity. For critical applications, a trial run to assess the selected viscosity may be necessary. A quick reference to the factors influencing the selection of viscosity is summarized in fig. 2–8.

DENSITY

It is the simplest property and is perhaps not considered an important one. But its relation to important properties like kinematic viscosity and compressibility makes it as an important parameter for designers. The higher the density of the oil, the higher the power required in the pump for a given flow rate. The value of density also affects the oil inlet conditions. Density also influences the filtration, as a higher density fluid will keep many particles in suspension. Density does not vary significantly with temperature. It decreases with an increase in temperature, and increases with a fall in temperature. But it increases with increasing viscosity. Low viscosity index (LVI) oils are denser than MVI, which are denser than HVI oils. So density of paraffinic base oils is lower compared with other type of mineral oils. Among hydraulic fluids used in industry, the density of mineral hydraulic oil is the lowest.

Currently the international practice is to quote the density at 15°C (59°F). The density of hydraulic oil is around 0.88. It is expressed in gram/per cubic centimeter. Readers may note that common terms such as light oil or heavy oil have nothing to do with density; they refers only to low viscosity and high viscosity oil. Often water is referred to as a liquid that is lighter than oil—it refers to viscosity only, not density. In fact, water can be called heavier than oil, as water is denser than oil.

BULK MODULUS

Generally in the design of many hydraulic systems, hydraulic oil is assumed to be incompressible. This is reasonably accurate as long as the operating pressure is moderate, say, 30 bar (450 psi) or less. But above this pressure, compressibility of the oil has to be taken into account in design calculations. Bulk modulus, or compressibility, is the measure of resistance to reduction in volume of fluid under pressure, and it is similar to Young's modulus in solids. It is the reciprocal of compressibility. Bulk modulus is related to the stiffness of the fluid. The higher the bulk modulus, the stiffer the system. Bulk modulus value increases with pressure but decreases with temperature, yet the effect of temperature on bulk modulus is more pronounced. The compressibility of oil, for example, at 100°C (212°F) is 25 to 30 percent more than at 20°C (68°F). Note that air entrapped in even a small quantity can drastically reduce the value of bulk modulus. For example, a volume of 1 percent air entrapped in the oil, say at a pressure of 35 bar (508 psi), can reduce the value of bulk modulus to as low as 25 percent of the normal value (air is 10,000 times more compressible than oil). At a higher pressure level, however, the reduction in bulk modulus is relatively less. Bulk modulus also varies slightly with the viscosity of the oil—the higher the viscosity, the higher the bulk modulus.

Bulk modulus is an inherent property of the oil and cannot be improved by means of additives. Naphthenic oils, for instance, are more compressible than paraffinic oils. As a general guide, the compressibility of oil can be taken as about 0.5 percent per 70 bar (1015 psi) up to 345 bar (5000 psi) for temperatures up to 40°C (104°F). The unit of bulk modulus is the same as that of pressure.

Significance of Bulk Modulus

Bulk modulus is important in many practical ways. For example, the dynamic performance of the hydraulic servo system is very dependent on this property. A lower value of bulk modulus due to air entrainment, in a servo system, for example, may lower the natural frequency of hydraulic actuators, which causes instability in the system. The response of a control system is also dependent on this parameter. The input and output motions are not rigidly coupled (fluid is the coupler), but there is a dynamic relationship similar to that of a mass-spring system. A high rate of response means that at the point of pressure change, the filling or emptying of hydraulic lines should be faster so that time lag between input flow and output motion is kept to the minimum. It can be seen that the response and stability characteristics of a hydraulic control system are very much influenced by fluid compressibility.

Bulk modulus is also a significant parameter in deciding the performance characteristics of machinery. In machine tools, for instance, excessive elasticity or compressibility of the fluid can impair the positioning accuracy of slides, thereby affecting the dimensional control on the workpiece to be machined. Also, in precision machine tools, such as grinding machines, shock waves in the fluid can be witnessed during the table reversals. In a poorly designed system, shock during the reversals of the table will have an adverse effect on the quality of the ground surface. This is a critical parameter where a high volume of fluid or high accuracy is involved. In hydraulic presses, when valve ports are opened suddenly due to a sudden release of compressed energy, pressure pulsation may take place. This results in shock waves in the fluid, called water hammer, as the volume of compressed fluid involved is quite large. The sudden release of pressure inside the ram creates flow rates of very high velocity, leading to the phenomenon of the water hammer, which can cause fractures in the machine structures. The effect associated with such a sudden and sharp change in pressure is taken care of by incorporating an exhaust valve with decompression features. In all these cases, the compressibility of the fluid should be carefully taken into account in all design calculations in order to achieve the desired performance level as well as to avoid any malfunctioning of the system.

Classification of Bulk Modulus

Bulk modulus is broadly expressed at isothermal and isentropic conditions, and fig. 2–9 explains the different ways of defining bulk modulus. Isothermal bulk modulus refers to the value at constant temperature conditions and is often referred to as static value. This value is applicable to conditions where changes

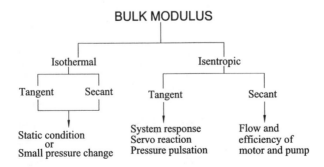

FIGURE 2–9 Classification of Bulk Modulus

of pressure and volume are slow, for example during the pressure buildup in a slow-acting hydraulic ram. Without any qualification, bulk modulus normally refers to the isothermal tangent bulk modulus.

When the hydraulic system operates at high speed, which is common in most cases, the heat generated due to compression is not dissipated, and the compression of the fluid is said to be isentropic. In other words, the isentropic bulk modulus is under the condition of constant entropy and is referred to as dynamic value, as this situation occurs where pressure changes are rapid. It is an important parameter to a designer in two ways. First, when the change of volume (compression or decompression) occurs in a high-speed pump, the value of this bulk modulus is used. Second, as mentioned earlier, it is used in the calculation of system response.

As given in fig. 2–10a and 2–10b, the isothermal and isentropic values are again expressed as tangent and secant bulk modulus. The tangent bulk modulus (fig. 2–10b) is related to the actual slope of the compression curve at any point, i.e., the true rate of change at working pressure, and is used for the calculation

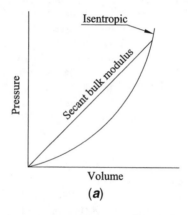

FIGURE 2–10a   Secant Bulk Modulus

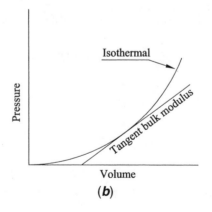

FIGURE 2–10b   Tangent Bulk Modulus

of system response, servo reactions, pressure pulsation, etc. The secant bulk modulus (fig. 2–10a) is the ratio of total change in pressure to the total change in volume, i.e., from atmospheric to working pressure, and is proportional to the slope of the line connecting a point on the PV curve to the origin. This value is used in the calculation of flow and efficiency of the pump and motor. Readers may note that the curve in the figure indicates that it is difficult to compress the oil at high pressures—the volume decrease is much less for an increase in pressure beyond a certain point, and the value of bulk modulus varies with the pressure at which the measurement is made. Mathematically, bulk modulus is expressed in the following way in accordance with the ISO convention [26].

Isentropic tangent bulk modulus (K_S) is given by the relation:

$$K_S = -V(\partial p/\partial V)_S$$

Isentropic secant bulk modulus (B_S) is given by the relation:

$$B_S = -V_0[(p - p_0)/(V_0 - V)]_S$$

Where ∂p = change in pressure
         ∂V = change in volume
          p = gage pressure
        p_0 = atmospheric pressure (i.e., zero gage)
         V = volume at pressure p
        V_o = initial volume at atmospheric pressure and temperature
subscript s = refers to constant entropy.

The same relation holds good also for isothermal secant and tangent bulk modulus, except that the values are taken at the isothermal (constant temperature) condition instead of the constant entropy condition. The negative sign indicates that the tangent slopes are negative, and in practice they are usually omitted while quoting values. At low pressures, in all calculations the difference between the two values can be ignored. However, at a high temperature or in a high pressure system the difference in compressibility data can upset the calculations. The value of bulk modulus for a given pressure and temperature can be graphically obtained from the procedure given in the ISO standard 6073:1997 [26].

FOAMING RESISTANCE

Foaming resistance is another important property of the oil, and it governs many performance factors of a hydraulic system. Air gets entrained into the hydraulic system in so many ways:

• Air bubbles get mixed with the oil streams returning from the system to the tank. High velocity of the oil and oil falling from a greater height in the tank can aggravate this problem.

FIGURE 2–11   Cavitation Erosion on the Vane Pump Pressure Plate (Courtesy: Yuken India Limited, Bangalore)

- Air leakage through loose fittings on the suction lines, low fluid level in the reservoir, and air leakage through worn seals, for example in cylinders.
- All hydraulic oils contain about 6 to 8 percent dissolved air by volume. When the suction pressure falls to a very low level, i.e., lower than the level of vapor pressure (this happens in a poorly designed suction line), the dissolved air may be released from the oil and cause foaming and cavitation.

Many studies have been conducted on foaming and air entrainment [29–37]. The effect of entrained air in the low pressure system is more critical than in the high pressure system, as air is dissolved easily under high pressure. As mentioned earlier, even a small amount of entrained air can drastically reduce the value of bulk modulus, thereby reducing the stiffness of the hydraulic system. Such a condition will lead to jerky movement and poor system response. Also, if the air bubbles do not break down quickly when they migrate to the surface, foaming will occur. This can lead to problems. For example, if air is drawn along with the oil into the pump, the volumetric efficiency of the pump will fall, and this continuing action can lead to a possible breakdown of the pump itself.

It's important to keep the pressure drop in the suction side as low as possible and to ensure that there is always a net positive suction head. And it is essential that the pump should have sufficient suction to overcome the pressure drop in the line. The fluid velocity in the suction lines should not exceed 1.2 m/sec (4 ft/sec) in order to minimize the pressure loss in the suction side. Undersized

FIGURE 2–12 Damage to the Vane Pump Support Plate due to Cavitation (Courtesy: Eaton Corporation)

suction lines and strainers, too many bends or elbows, a very long suction pipe, too many pipe joints, and clogged strainers are some of the causes of undesirable vacuum conditions in the suction side that lead to the release of dissolved air in the oil. The released air from the oil will form air bubbles. As the oil and the air bubbles move from the suction side to the pressure side of the pump, the air bubbles will collapse. Collapse of the air bubbles can cause significant erosion to the metal surface, and this phenomenon is known as gaseous cavitation. The metal erosion formed in a vane pump pressure plate due to cavitation is shown in fig. 2–11. The high negative pressure developed in the suction side due to the clogging of the suction strainer was the cause for this cavitation—a simple thing, but it can lead to such costly damage! Users should insist that the strainer be placed in an easily accessible position for cleaning. In vane pumps, large bubbles can cause the vane to collapse and pound (repeated hitting of the vane with heavy force), and this condition can lead to rapid wear of the vane and ring. The pump can lose its life within an hour under such conditions [38]. Another cavitation failure to the vane pump support plate is shown in Fig. 2–12.

Hence considerable attention should be given to the design of suction lines in order to keep the pressure drop as low as possible. Also, suction inlet lines should not become a source for air entrainment. For example, suction lines should be immersed deeply into the reservoir so that the suction pipe is always below the lowest possible oil level in the reservoir. And discharge lines from the relief valve should be kept below the oil level to avoid any air being carried to the oil.

Foaming also accelerates the oxidation of the oil. When entrained air bubbles pass from atmospheric pressure to pump pressure, the temperature can rise to as high a value as $1,000°C$ ($1,832°F$), resulting in thermal cracking and nitration of the oil [58], which leads to formation of insoluble deposits. High operating pressure can aggravate this problem. As the oil absorbs the heat of decompression, this can also raise the temperature of the oil. If the condition remains unchecked, degradation of the oil will be faster than normal.

To mitigate the foaming effect, hydraulic oils are fortified with antifoam additives as dispersion, the droplet size not exceeding 10 microns in diameter. The additive is *dispersed* not *dissolved* in the oil; in the dissolved condition, the additive becomes ineffective. This also brings us to the fact that the additive used should have a lower surface tension than that of the oil. In fact, sometimes the surface tension of the used oils is measured to determine the depletion of the antifoam additives. Low surface tension in the used oil is an indication that the oil has been depleted with antifoam additives. The more commonly used antifoam additives are liquid silicones, which are used in small concentrations of 1 to 10 ppm. These additives are absorbed in the very thin oil films between air bubbles in the foam; they render the oil foams inhomogeneous and unstable, and cause the film and the foam to break. Low viscosity and high surface tension of the oil also reduce the foaming tendency.

The presence of antifoam agents does not ensure against all foaming problems. In the case of persistent foaming, all causes should be investigated and rectified. Contamination such as water, grease, and particulates also increase the foaming tendency of oil. This emphasizes the importance of the golden rule "Keep the oil clean." Venting valves should be provided at appropriate places so that air entrapped in the system can be bled off periodically. Another factor that influences foaming is residence time (tank volume divided by the flow rate of the pump). If the oil passes from the drain area to the pump suction area without adequate residence time, i.e., the rate of oil circulation is too great, sufficient time may not be available for the foam to collapse. Such a situation will lead to persistent foaming. This problem can be remedied only by increasing the residence time. Residence time bears a direct relation to the degree of aeration. Depending upon the size and design of the system, residence time may vary between 2 minutes and 3 minutes, though 5 minutes is also used in some extreme cases. If the system is designed, for example, with fixed displacement pumps, the residence time must be longer, as the oil is more stressed. On the contrary, with variable delivery pumps the oil does not work that hard, hence residence time can be relatively shorter. Persistent foaming cannot be solved by further addition of any antifoam agents; you must remove the causes or switch to another brand of oil. The foaming property of oil is summarized in fig. 2–13.

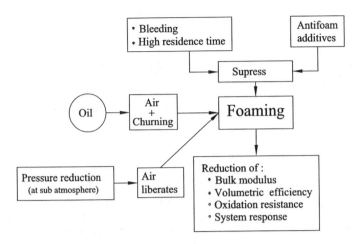

FIGURE 2–13 Foaming Property of Oil

AIR RELEASE PROPERTY

Air release property is closely related to antifoam performance. Foaming is a surface phenomenon, in the sense that foaming is thought of as a large amount of gas enclosed in a thin film of oil, which forms a low oil content structure in the surface of the oil [31]. On the other hand, air entrainment is the dispersion of a small amount of air as tiny air bubbles throughout the fluid, and the bubbles are separated by a relatively thick oil film. Air entrainment in the system mostly escapes to the surface as foam. However, the balance of air, if any, particularly in a high pressure system, is entrained in the oil, which results in spongy controls, loss of efficiency, cavitation, noisy operation, pressure spikes, temperature increase, and accelerated oxidation. It is pertinent to note that per the familiar Henry's law, the solubility of gas in a liquid is directly proportional to the absolute pressures; hence this property is more significant where the pump works under high pressure with fast pressure cycle changes. For such applications, oil with a good aeration property should be used. Under extreme conditions, air entrainment can cause air-locking in pumps. Rapid release of air is therefore an important property of hydraulic oil. The ability of the oil to expel dispersed air is known as its air release property.

Aeration in the system is often characterized by the pump whining—a distinct, long, high-pitched sound emanating from the pump that be easily heard. This noise ceases once the air entry is closed and the air is purged from the system.

The factors that contribute to air entrainment and air release should be understood clearly so that appropriate preventive action can be initiated. Viscosity, system pressure, contaminants, and additives affect the rate at which air is released from the oil. For example, air release time increases with increase in viscosity and also in highly contaminated oil. Low temperature in the reservoir reduces the rate of bubble collapse, whereas a high temperature promotes bubble formation but releases air more rapidly.

Leaking oil seals in the pump and leaking joints in the suction lines are potential sources for air entrainment. Besides that, low oil level, a high pump speed, and a return line above the fluid level can also let air into the system. Apart from eliminating these causes, care in the reservoir design coupled with good housekeeping of the fluid will also considerably lessen this problem.

OXIDATION STABILITY

As the life of the oil is dependent on its ability to resist oxidation, this is an important property from the user's point of view. In economic parlance, it is the property that decides the dollar value of the oil. When oxygen combines with hydrocarbon molecules, a chain reaction takes place, producing soluble as well as insoluble substances that lead to degradation of oil. As oxidation progresses, the oil becomes dark and acidic, and forms sludge and varnish. Oxidation also increases the viscosity of the oil and reduces the foaming resistance and air release property. Sludge may deposit in valve ports and pipelines, restricting the oil flow and impairing the smooth functioning of the system. Thus the oil becomes nonserviceable.

The chain reaction due to oxidation takes place in the following way:

Chain initiation: RH (hydrocarbon) = R· (free radical) + H· (hydrogen)

Chain propagation: R· + O_2 → ROO· (peroxy radical)

ROO· + RH → ROOH(hydroperoxide) + R·

Termination: R· + R· → R· R

The above sequence suggests that there are three stages in the oxidation process. First there is the generation of free radicals under the influence of heat and other catalysts. Next comes a propagation stage in which these free radicals react with oxygen and the oil to form hydroperoxides and other free radicals in a chain reaction. The last phase is termination, in which the radicals combine with oxidation inhibitors. Typically this sequence may occur about five times before the chain is broken [39]. The decomposition of the hydroperoxide leads to the formation of alcohols, keytones, aldehydes, and organic acids, which form sludge and sticky substances.

Parraffinic hydraulic oils by themselves are more resistant to oxidation than straight mineral oils. However, all hydraulic oils today are fortified with an antioxidant additive to extend the life of the oil. Antioxidants are compounds such as amines, phenols, or zinc dialkyldithiophosphate (ZDDP) added in a small concentration to reduce the rate of oxidation. The antioxidant additives of phenolic types perform their function by breaking the chain and acting as free radical scavengers, whereas ZDDP types destroy peroxides.

As stated earlier, oxidized oil will have poor foaming resistance and air release properties, which can badly damage a machine. This was evident in a special purpose boring machine I observed. The machine was jerking and giving inconsistent feed, which was affecting the surface finish; the customer was very particular about the surface finish of the workpiece. As the oil was very badly oxidized, gaseous formation led to sponginess in the oil, which made the feed erratic. This problem was acute in the morning, when the oil temperature was low. Low temperature and the consequent high viscosity coupled with the high contamination of oil significantly reduces the rate of bubble collapse. Once the hydraulic system was cleaned and the oil was changed, the machine was restored to normal functioning. The problem was traced to the badly degraded oil after a long elimination of other causes, which of course cost a lot of time and money. Systematic fluid maintenance would have avoided the problem and the breakdown cost associated with it.

Effects of High Temperature

The main factors that contribute to oxidation are temperature, presence of water, and metal catalysts like copper, iron, and lead wear particles. Among these, temperature is the prime accelerator of oil oxidation. For every 10°C (18°F) increase in temperature, the rate of oxidation is doubled. A fully oxidation inhibited HVI oil, for example, might have a useful life of about 100,000 hours at 40°C (104°F), even when aeration and catalysts are severe. When the operating temperature is raised to 60°C (140°F), the life of the oil drops to 10,000 hours [39].

As the oil circulates from the pump to the sump, it is heated due to the churning and shearing action of the oil. Frequent dumping of the oil through relief valves, internal leakage in the pump, rapid compression of entrained air, inadequate tank capacity, and unwanted pressure loss are some of the sources through which heat is added to the oil. High velocity of the fluid in the lines, for instance, is one of the main causes of unwanted pressure loss in the system. The recommended fluid velocity in the pressure lines is around 3 to 4.5 m/sec (10 to 15 ft/sec) for the low pressure system—about 35 bar (500 psi)—and for the high pressure system—70 to 200 bar (1000 psi-3000 psi)—it should be limited to a maximum of 6 m/sec (20 ft/sec). In the return lines it should not be more than 4 m/sec (13 ft/sec). Pressure and return lines should be sized taking these recommended fluid velocities into consideration so that pressure loss can be kept to the minimum. A clear understanding of the sources of heat generation could go a long way in controlling the bulk temperature of the oil. Incorporation of a suitable heat exchanger should always be considered where bulk oil temperature is high, say 50 to 60°C (122 to 140°F), and where performance requirements are also high.

Besides accelerating the oxidization process, high temperature of the oil can also cause costly damages; one such scenario is illustrated in fig. 2–14. This cutting of corners may lead designers to avoid incorporating a heat exchanger in the system, even when it has to operate under high ambient temperatures of 40°C (104°F) or more. As an alternative to the heat exchanger, a perceived

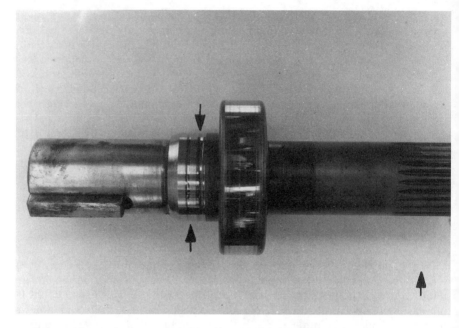

FIGURE 2–14 Grooves Cut by the Oil Seal on the Pump Shaft Due to High Temperature (Courtesy: Yuken India Limited, Bangalore)

cost-effective alternative is to increase the reservoir size to five times the pump flow rate. This is a poor alternative because a large tank does not control temperature and also reduces the effectiveness of filtration.

The damage to the pump shaft due to high temperature of the oil as shown in fig. 2–14 was caused by this false economy. The bulk temperature of the oil was more than 60°C (140°F), and obviously the temperature inside the pump was much more than the bulk oil temperature. Normal pumps are not meant to operate at such elevated temperatures, particularly on continuous duty. As we know, normal engineering materials expand due to heat, while seal elastomers shrink, become hard, and can even cut deep grooves into the hardened shaft. As a result, the pump failed. In such operating systems, a heat exchanger should be considered an essential element, not construed as an unnecessary added cost.

In another case, with a view to economizing on the cost of a heat exchanger in an automatic lathe, the hydraulic oil reservoir was kept inside the coolant tank, as shown in the fig. 2–15. When the turret indexing malfunctioned, the problem was traced to water contamination in the oil. As unskilled people usually do the topping-up, by mistake the coolant was poured into the hydraulic reservoir! As the user had three machines of the same type, in order to avoid further human errors the user modified the system by separating the two reservoirs and adding a heat exchanger to the hydraulic system. In addition, the two reservoirs were given a distinct identity by color, shape, and labeling. This incident underlines the fact that there are no shortcut solutions to any problems. The

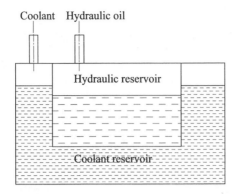

FIGURE 2–15 Coolant Reservoir Encasing Hydraulic Reservoir

system designed should have a sound engineering basis with a touch of practical consideration and should not be vulnerable to common human errors. Due to the traditional budget-oriented approach, many people are often sensitive to a higher initial cost of equipment and tend to look upon many essential items as extra expenses. The long-term benefits of a well-engineered system are seldom appreciated, and the extra money that is spent later to repair the equipment is ignored. The modification cost incurred by the user is much higher than the cost of a heat exchanger as an OEM item.

Water and metal contaminants are also important components for oxidation. Copper, for instance, in a concentration of only a few parts per million is sufficient to hasten the process of oxidation. Traces of copper can easily get into the system by the action of the oxidation products of copper pipes. A well-installed and well-maintained filtration system is essential to keep the oil free of contaminants, which controls the oxidation process. Here are a few tips [39], to reduce the oxidation of oil:

- Keep the bulk oil temperature as low as possible.
- Avoid excessive agitation that will cause foaming, which makes air readily available.
- Do not use metals like copper (copper tubes, for example), as copper has a high catalytic activity.
- Ensure that water does not enter the system.
- Keep contaminants out through good filtration.
- Remove all the old oxidized oil when an oil change is made, as partially oxidized oil is an effective catalyst for further oxidation.

As the oxidized oil becomes acidic, the level of acidity in the oil is used to indicate the oxidation level. Acidity level is termed as the Total Acidity Number (TAN) or the Neutralization Number (NN) and is expressed in mg KOH / g—the number of milligrams of potassium hydroxide required to neutralize the acidity of one gram of oxidized oil. It is generally accepted that when TAN value reaches 2, it's time to change the oil. TAN value can be determined per the procedure

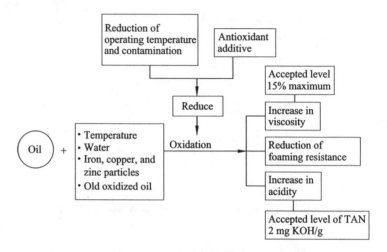

FIGURE 2-16 Oxidation Property of Oil

given in ASTM standard D 974 or IP 139 (next chapter). Fig. 2–16 summarizes the oxidation properties of the oil.

ANTIWEAR PERFORMANCE

One of the basic functions of any lubricant is to separate the sliding surfaces, which are set in motion by a fluid film. However, in heavily loaded contacts encountered in equipment such as gear pumps or vane pumps, it may not be possible to obtain the full fluid film conditions at all times. Most of the time these contacts operate under a mixed lubrication condition in which oil film does exist between the surfaces, but the film is ruptured intermittently by the surface asperities, promoting considerable metal-to-metal contact. This condition can accelerate the wear of the pump components, thus reducing the life of the pump. As pump operating pressures and speeds increased, the antiwear performance of the fluid became important. Antiwear performance of hydraulic oil is enhanced by the addition of antiwear or load-carrying additives.

Antiwear additives work by reacting with metal asperities to form protective films, which help the oil film reduce intermetallic contact and wear. For example, when zinc dialkyldithiophosphate (ZDDP) is used as an antiwear additive, it reacts with the metal surface and forms metal sulphide and thiophosphate layers. These layers reduce the friction and wear when sliding takes place under boundary lubrication conditions. Thus the protective film helps the parts to slide with less friction and a minimum of metal loss. The action of antiwear additives in any lubricant is a complex subject [40–55]. The quality of the base oil used also can have a significant influence on the antiwear performance of the oil—the thicker the base oil, the better the antiwear performance. It may be also noted that the antiwear performance of oil becomes important when the pump size increases as the sliding velocity of the vane tip increases.

A Note on Antiwear Additive

An important property of mineral oil that many may not be aware of is its chemical inertness. In other words, it is compatible with a wide variety of materials. In the last few decades, the focus of lubrication technology was mainly on the development of additives—chemically active materials. This has led to the belief that oil with many additives performs better than oil with fewer additives. Additives are like antibiotics—besides benefits, they can produce side effects and create other problems. Users should exercise discretion in selecting an oil with many additives, and antiwear additive is not an exception.

Since the 1940s the only hydraulic oil used was R&O oil (rust and oxidation inhibited oil). But this oil could not cope with the needs of vane pumps operating above 70 bar, and wear problems became more pronounced. Tricresyl phosphate (TCP) as an antiwear additive was added to R&O oil to enhance the antiwear performance of hydraulic oil. But in the early 1960s its inferior performance in relation to the antiwear additive used in motor oils, zinc dialkyldithiophosphate (ZDDP), came into focus. Hydraulic oils blended with ZDDP additives then came onto the market. It was used extensively because it served the dual purpose of being an antioxidant and an antiwear additive. Though oil with ZDDP was successful in that regard, it also created two critical problems [40,41]:

- **Incompatibility**—Oil containing ZDDP reacted with silver plating, bronze components, and white metals, causing corrosive wear.
- **Poor filterability**—ZDDP combining with moisture decomposes into various metal particulates and leads to filter plugging (even 10 μm rating filter). Many servo systems demand filtration to the level of 5 μm or less; oil with ZDDP additive could not cope with the demand of such applications. Also, deposits from ZDDP can cause stickiness in valve movement, impairing the precision of the hydraulic system.

The shortcomings of ZDDP oil were overcome by the formulation of ashless oil (sulfur-phosphorous type) and a more stabilized zinc-based oil [41–43,50]. These new formulations, it is claimed, meet even severe hydraulic pump requirements and are suitable for both vane and piston pumps.

Users can still bank on R&O oils for many applications and select an antiwear type only when it is strictly warranted by pump operating conditions so that compatibility problems can be lessened.

FILTERABILITY

The higher demands placed on the performance of hydraulic systems mean that hydraulic components operate under much finer clearance than before. For example, typical clearance at the vane tip or in the piston pump is as low as 0.5 μm, and at seals it can go further down to 0.05 μm. This emphasizes the fact that besides good filtration, oil should possess good filterability. Filterability of hydraulic fluids is a measure of the ease with which the fluid may pass through a filter. It also indicates the speed at which the filter can get blocked.

There is now a growing awareness that poor filterability of oil can cause costly breakdown of machinery. Particularly in servo systems, filtration of oil to a fine degree is a prime requirement. Composition, thermal and hydrolytic stability of the additives, and homogeneity of the fluid are a few factors that greatly influence the filterability of the fluid.

The most widely used filterability test is the Denison filterability test procedure TP 02100 method to evaluate filter plugging. It is widely recognized by fluid manufacturers as a useful test. Other methods are AFNOR and ISO. The details of these test procedures are given in the next chapter.

POUR POINT

Pour point property characterizes a fluid's ability to flow at low temperatures. Pour point of a hydraulic fluid is the lowest temperature at which the fluid ceases to flow. This property is significant in cryogenic or refrigeration systems but is seldom a matter of concern in most industrial hydraulics, except in a situation where the hydraulic system has to operate in a low ambient temperature. It is important to recognize that fluid viscosity is more critical in terms of flow to the pump inlet than pour point since viscosity will be much higher at lower temperatures. The fluid viscosity at temperatures above the pour point should be within the pump start-up viscosity limits.

Pour point is closely related to type of base oil, i.e., the type of hydrocarbon molecule and its molecular weight. For example, naphthenic oils have low pour points, whereas paraffinic ones of the same viscosity grade have a much higher value. This is due to the precipitation of wax in the paraffinic oil. At low temperatures, the wax crystals get separated out in the form of needles and platelets and form a network, thus preventing the still liquid components from flowing. Pour point depressants can lower the pour point of a fluid by as much as 30°C (86°F). These additives change the size and shape of wax crystals from needles and platelets to more spherical crystals. These changed crystals are less likely to form networks and thus allow the oil to flow. Some commonly used pour point depressants are polymethacrylates, styrene esters, and crosslinked wax alkylated phenols. The pour point of mineral hydraulic oil is generally around −20°C (−4°F), which is much more than normally required in industrial hydraulics.

RUST AND CORROSION INHIBITION

Preventing water getting into the hydraulic system in all situations may not be an easily achievable task, and ingress of even traces of water can cause rust problems. Also, heteroatoms like oxygen, sulfur, or chlorine can react with the metals and form corrosive compounds. Though mineral oil by itself possesses antirust properties, it is augmented by polar compounds, which adsorb onto the metal surfaces; this is often referred to as "plate out" and together with the oil it forms a thin film that is impervious to water and other heteroatoms. Thus metal surfaces are insulated from coming into contact with corrosion-inducing elements, and rust formation is prevented. Additives such as alkylamines, metal sulfonates, or salts of fatty acids in small amounts, usually 0.5% or less, are added to the oil to provide rust and corrosion inhibition.

COMPATIBILITY

Mineral oils by themselves are inert and compatible with a wide range of materials, but this is not so when they are blended with additives. Flexible sealing elements made from many types of elastomers are indispensable to hydraulic systems. Similarly, a wide variety of metals are used in hydraulics. Users should be aware of the effects the hydraulic fluid may have on elastomers or metals used in the hydraulic system. Compatibility of hydraulic fluids with various materials has been recognized as an important property in recent years, particularly after the incompatibility of ZDDP on certain materials became known. Yet there is no accepted test method for evaluating the metal compatibility of oil, though some manufacturers use the Cincinnati Machine's [50,51] thermal stability test procedure (Chapter 3). Test procedures are available, however, to evaluate the compatibility of the fluid with respect to seals.

Compatibility with Seals

Readers are aware that hydraulic fluids contain a host of chemicals as additives; seals also contain additives, for example, vulcanization agents, plasticizers, antioxidants, fillers, and accelerators [56]. This can lead to chemical incompatibility between the fluid and elastomer, which results in seal swelling, softening, or formation of insoluble products. The most common source of trouble in seals is absorption of fluid by the elastomer and its consequent swelling. Absorption of fluid causes plasticizer to leach out, and in some cases the effect is a net loss, resulting in shrinkage of the seal. Swelling of seals is always associated with softening, and shrinkage with hardening. Such effect on seals can cause leakage, internal pressure losses, temperature rise, and possible transmission malfunction. Seals may not be compatible with other lubricants as well. For instance, Fainman and Hiltner report [56] on the incompatibility of common greases that are used for the lubrication of O-rings.

Mechanical engineers, who are the principal users of hydraulic fluids, are hardly aware of the intricate chemistry of fluids and seals or their dependence on each other. Added to that, under the pretext of being "proprietary", the composition of fluids or seals is seldom divulged to users. In these circumstances, designers find it difficult to give full consideration to compatibility of fluids, seals, and the material used in the system for the given application. Though available standard tests on seal compatibility may mitigate this problem, for critical applications it is advisable that users devise their own test method for assessing the seal compatibility for a specific application in order to prevent costly equipment failure. For example, Murray has devised his own test methods to evaluate seal compatibility for military applications [57].

For a long time, the aniline point was used to indicate the compatibility of seals. The aniline point is the lowest temperature at which oil is completely miscible with aniline in equal volume; it is expressed in degrees centigrade. Low aniline point means high swelling and vice versa. A low aniline point is an indicator that oil contains a higher content of aromatic hydrocarbons, which cause the elastomer to soften and swell. The test has no direct bearing on any seal materials; its correlation with actual swelling is not satisfactory and hence it has lost much of its validity as an indicator of compatibility of seals. Currently the compatibility of seals to hydraulic fluids is evaluated per the test procedure prescribed in ISO 6072 (Chapter 3).

DEMULSIBILITY

Demulsibility is a fluid's ability to separate itself from water quickly and completely. Water or moisture ingressed into the system condenses above the oil level during idle periods and slowly settles at the bottom of the tank. During operation, churning causes water to mix with oil, which forms emulsion. Emulsified oil can interfere with the smooth operation of valves and pumps. A clean system and a well-refined oil will quickly separate the water from emulsion. Contaminants and oxidized particles will considerably impair the demulsibility characteristic of the oil. This implies that keeping the oil clean can eliminate many problems; otherwise, one problem may either lead to or accentuate other problems. Oil that has high oxidation resistance retains water-separating ability over a long period. Apart from protecting the system from water, in a large system water drain-off is provided in the reservoir to control the emulsification of oil. The commonly used demulsifiers are polyethylene glycols or polyoxy-alkylated phenols.

HYDROLYTIC STABILITY

Hydrolytic stability is another property relating to the effect of water on hydraulic oil. Hydraulic systems can get contaminated with water in so many ways—for example, in a machine tool where the cutting fluid is water-based. Hydraulic fluid should remain chemically unchanged or stable in the presence of contaminated water; otherwise insoluble products or corrosive materials can cause filter plugging. Hydrolytic instability can also cause corrosion in bronze parts commonly used in piston pumps. Any premium-grade hydraulic oil should have resistance against hydrolytic action. In essence, this property focuses on the chemical stability of the additives used in the oil in the presence of water. The instability of the antiwear additive ZDDP in the presence of water is a classic example of hydrolytic instability.

THERMAL STABILITY

Per the ASTM definition, thermal stability is the resistance to permanent changes caused solely by heat in the properties of the fluid. This definition is now expanded to include the effects of catalysts such as copper and iron normally present in hydraulic fluids. In other words, this property focuses on the chemical stability of oil at elevated temperature in the presence of metal cata-lysts. This problem is more pronounced where the reservoir size is small, i.e., the residence time is small, as the bulk temperature of the oil will be high.

In recent years this property has gained more importance. In fact, many leading oil manufacturers make it a point that their product passes the ther-mal stability test. Severe operating conditions such as high speed and load can cause temperature rise; this is more pronounced in hot spots, which can cause thermal decomposition of certain additives. These thermally decomposed par-ticles can react with yellow metals and their alloys, causing corrosion, forming sludge, and increasing viscosity. As a result, filter plugging and deterioration in system cleanliness can occur, which in turn cause major system failure. Periodic analysis of targeted oil samples has confirmed that oil does exhibit ther-mal instability [58]. The accepted test procedure to evaluate this property is Cincinnati Machine's thermal stability test, which is described in the next chap-ter. It is a very demanding test; not all oils can pass it. There is also an ASTM standard similar to this procedure.

CONCLUSION

Acquiring a comprehensive knowledge of all these properties of hydraulic fluids is essential to appreciating what a fluid can do and what it cannot do. Assimi-lation of this knowledge could culminate in the selection of the right fluid for a given application, as well as better care and application of hydraulic fluids. This knowledge can also help reduce communication barriers between fluid and equipment suppliers, which can result in the right kind of interaction. Mis-applications, system malfunctioning, and costly mistakes can be considerably minimized, if not altogether avoided. Also, more important, this expertise can provide the necessary theoretical corrections for troubleshooting. In sum, it can save significant amounts of time and money, and help run the hydraulic system to the best efficiency.

Chapter 3

TEST METHODS ON HYDRAULIC FLUIDS

The various properties of hydraulic fluids outlined in the previous chapter are qualified and quantified by standard tests. Along with many other institutions, the American Society for Testing and Materials (ASTM), the Deutsches Institut fur Normung (DIN) and the Institute of Petroleum (IP), U.K., did pioneering work in formulating test standards for petroleum products. This has helped lubricant manufacturers as well as users to maintain and enhance the quality of lubricants over the years. If it weren't for the test standards, the performance of hydraulic fluids or lubricants would not have reached today's high level. For readers, the tests described here will give a better insight into the technology of making high-performance hydraulic fluids, and will help them understand the various specifications of hydraulic oils. Also, when confronted with a problem, this knowledge will help readers identify the kind of tests to be carried out in order to diagnose and solve the problem. Formulating test standards is a dynamic process, and as such they are always under periodic scrutiny and are revised when necessary. Not only old standards are revised or withdrawn, but new performance criteria are also added.

The test procedures described here are based on ASTM standards [1] unless otherwise specified. Tests are always conducted under specific conditions, which usually involve elaborate preparation of test apparatus and test samples. Those are not described here and readers may refer to the relevant standard for details. The list of test standards and the numbers for various national standards are given in Table 3–1. The table gives only the equivalent standards; these standards are not necessarily identical to one another, and there could be variations in the test procedure and test conditions from one national standard to the next. Hence, when comparisons are made, caution must be exercised—test parameters, test conditions, and methods of measurement must be taken into account.

COLOR (D 1500)

Oil displays a characteristic color in reflected light, and the transmitted light will reflect the true color. For example, paraffinic oils tend to be green, whereas

Table 3–1   Equivalent Standards for Testing Hydraulic Fluid Properties

Sl. No.	Properties	ISO	ASTM	DIN	IP	IS: 1448	BS
1	Color	2049	D 1500	51578	196	P: 12	5859
2	Kinematic viscosity in cSt	3104	D 445	51561 51562	71	P: 25	2000 Pt. 1
3	Viscosity index	2909	D 2270	2909	226	P: 56	4459
4	Pour point	3016	D 97	3016	15	P: 10	2000 Pt. 15
5	Flash point and Fire point COC	2592	D 92	51376	36	P: 69	4689
6	Antirust	7120	D 665A D 665B	51585	135 A 135 B	P: 96	2000 Pt. 135
7	Copper corrosion test	2160	D 130	51759	154	P: 15	2000 Pt. 154
8	Demulsibility	6614	D 1401	51599	19	–	2000 Pt. 19
9	Antifoam	6247	D 892	51566	146	P: 67	2000 Pt. 146
10	Air release test	9120	D 3427	51381	313	P: 102	2000 Pt. 313
11	Vane pump wear test	–	D 2882	51389 Part 2	281	–	2000 Pt. 281
12	Oxidation stability	4263	D 943	51587	280	P: 106	2000 Pt. 280
13	Hydrolytic stability	15596	D 2619	–	–	–	–
14	Seal compatibility	6072	–	53521, 53505	278	–	4832
15	Water content by distillation (Dean and Stark method)	3733	D 95	3733	74	P: 40	4385
16	Neutralization number (TAN) (Color titration method)	6618	D 974	51558 Part 1	139	P: 2	2000 Pt. 139
17	FZG gear rig test	–	–	51354 Part 2	334	–	–
18	Thermal stability	–	D 2160 D 2070	–	–	–	–
19	Shear stability	–	D 5621 D 3945	51382	294	–	–

ISO–International Standards Organisation
ASTM–American Society for Testing and Materials
DIN–German Industrial Standard
IP–Institute of Petroleum, U.K.
IS–Indian Standard
BS–British Standard

naphthenic oils have a blue cast. The color of oil is not a parameter indicating the quality of the oil. Nonetheless, it does give an idea about the condition of the oil and gives rough guidance to the contamination level of the oil. It is not a test used for diagnosing oils-in-service. Manufacturers use color as a control parameter while refining and blending.

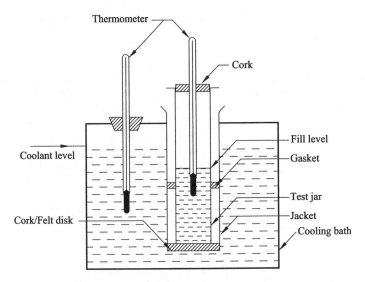

FIGURE 3-1 Apparatus for Pour Point Test

The sample oil contained in a glass jar is placed inside a colorimeter filled with distilled water. Then the light source is switched on, and the color of the sample is compared with ASTM standard color discs ranging in value between 0.5 and 8.0, in increments of 0.5. If an exact match is not found and the sample color falls between two standard colors, the higher of the two colors, i.e., darker color is reported. Generally, the color of lower viscosity hydraulic oils ranges between 1 and 1.5, while the color of higher viscosity oils is around 3, which also suggests that high viscosity oils are darker than low viscosity oils.

POUR POINT (D 97)

This test characterizes the flow characteristics of the lubricants at low temperature. The pour point of hydraulic oil or of any petroleum oil is the lowest temperature (expressed in multiples of 3°C) at which the oil will pour or flow when it is chilled, without any disturbance under definite prescribed conditions. After preliminary heating, the sample is cooled at a specified rate and examined at intervals of 3°C for flow characteristics. At each 3°C below the starting temperature, the test jar is removed and tilted just enough to ascertain the movement of the specimen in the jar. It is then further ascertained by holding the jar in a horizontal position for five seconds. If the specimen moves, the test is repeated until no movement is observed in the horizontal position for five seconds, and that temperature is recorded. To the recorded temperature, 3°C is added and reported as the pour point. Hydraulic oils having a pour point of –20°C (–4°F) are very common today, which is much higher than normal industrial requirements. Figure 3–1 illustrates the test apparatus used.

FLASH AND FIRE POINT BY CLEVELAND OPEN CUP (D 92)

Both flash and fire point are related to the fire resistance of the fluid, and hence they are not considered so relevant to the majority of common industrial applications. It is also generally felt that this test does not really simulate possible fire-hazard conditions. In fact, Burgoyne et al., reports that flash point is an arbitrary value and has no relation to any specific fire hazard [3]. However, the test qualitatively distinguishes one fluid's flammability from another's, and this data may be useful for handling and storage purposes. Despite its shortcomings, flash point is still used as an important parameter for fire resistant fluids.

Flash point is considered to be approximate to the lower limit of flammability, which is defined as the minimum vapor concentration required for flame propagation in a specified oxidant atmosphere [4]. Flash point signifies the degree and nature of volatile materials present in the fluid, whereas fire point signifies the fluid's susceptibility to support combustion. The test method described is applicable only for viscous material whose flash point is 79°C (175°F) and above. Because hydraulic fluids have a low volatility, their flash point is always above 79°C. In the Cleveland Open Cup (COC) apparatus (fig. 3–2), the oil sample is filled to a specified level in the test cup. Initially the oil is heated at a rate of 14°C to 17°C/minute (25°F to 30°F) and then decreased to 5°C to 6°C/minute, as the temperature reaches the anticipated flash point (the last 28°C). In the last 28°C, at every 2°C (5°F) rise in temperature, a test flame is passed across the cup for about one second. The lowest temperature at which application of the test flame causes the vapors above the surface of the liquid to ignite is taken as the flash point. When the fire point also has to be determined, the test is continued until the test flame causes the oil to ignite and burn for at least five seconds. The recommended safe practice is that the bulk temperature of the oil in store or operation should be kept below the flash point.

ANTIRUST TEST (D 665)

The antirust property of mineral oils is well-established, and this test is conducted commonly at the manufacturer's end. This test is applicable to any inhibited mineral oil; it evaluates the ability of both used and new hydraulic oils to prevent rusting of ferrous parts when the oil is contaminated with water. The test is conducted in two stages: the first is called Method A and uses distilled water; the second is called Method B and uses synthetic seawater. For the details on the preparation of synthetic seawater, readers may refer to the standard. Both tests are run under identical conditions. The test consists of stirring a mixture of 300 mL of test oil with 30 mL of distilled water, maintaining the temperature at 60°C (140°F) for 24 hours. A special cylindrical test specimen that is $\phi 12.7 \times$ 68 mm (0.5" × 2.677") long, made of Grade 10180 of specification A108 or BS 970: Pt.1, 070M20 steel, is completely immersed in the test fluid. At the end of 24 hours the specimen is removed, washed with solvent, and rated for rust. The oil is considered to pass if the steel specimen is completely free of any visible rust under normal light without any magnification. The test apparatus is illustrated in fig. 3–3.

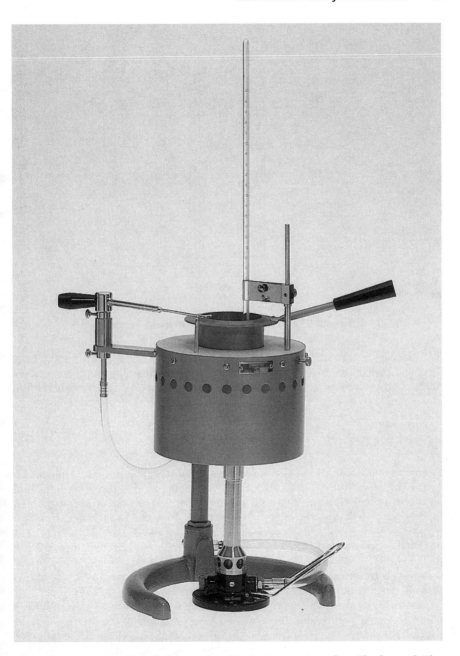

FIGURE 3–2 Cleveland Open Cup Test Apparatus for Flash and Fire
Points (Courtesy: Stanhope-Seta Limited, U.K.)

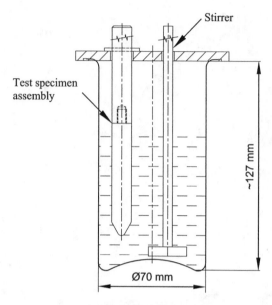

FIGURE 3–3 Apparatus for Antirust Testing

It is taken for granted that mineral-based hydraulic oils pass both these tests. This test procedure is also used to evaluate the antirust property of aqueous fluids such as water glycol and high water content fluid (HWCF), without adding any water.

COPPER CORROSION TEST (D 130)

The copper corrosion test is related to the antirust test and is applicable to petroleum-based lubricants and fuels. The purpose of this test is to determine how corrosive hydraulic oils and other petroleum products are to copper. Because copper-based alloys are used in many industrial machines, in piston pumps, for example, this test is an important one. A hard-tempered cold-polished copper strip that is 99.9 percent pure, 12.5 mm wide, 1.5 mm to 3 mm thick, and 70 mm long (1/2″ × 1/6 to 1/8″ × 3″) is immersed in 30 mL of sample oil, free of any water, at 100°C (212°F) for three hours. At the end of three hours, the copper strip is washed and compared with ASTM copper strip corrosion standards. There are four classes of corrosion per the standard. If the copper strip is only slightly tarnished, retaining almost the original color of the polished strip, i.e., light orange, the oil is said to pass Class 1. On the other end, Class 4 oils would turn the copper jet black or brown, signifying corrosion. Hydraulic oils marketed today generally pass the Class 1 classification. This procedure is also used for water-based fluids like HWCF.

In addition, this test is used to indicate the presence of any sulfur in the oil after refining. It can be used as well to determine whether the oil is contaminated with cutting oil, as sulfur/chlorine that is usually present in the cutting oil can cause corrosion.

DEMULSIBILITY CHARACTERISTIC OF HYDRAULIC OIL (D 1401)

The demulsification test was originally intended for turbine oils, but it is now used to measure the ability of water and any petroleum-based lubricants or synthetic fluids to separate from each other when mixed. The test is used for new oils as well as to monitor oils in-service.

A mixture of a 40 mL sample oil and 40 mL of distilled water is stirred at a speed of 1,500 rpm for five minutes at 54°C (129.2°F) in a graduated cylinder of 100 mL capacity. The time required for the separation of the emulsion thus formed is recorded. The limit for water separability is an emulsion reduction to 3 mL or less after standing for 30 minutes at 54°C. When this limit is exceeded, the volumes of oil, water, and emulsion remaining at are also recorded and reported. The separation time for the new generation hydraulic oils is generally between five and 15 minutes.

The test conditions are slightly modified for high viscosity oils. For oils of viscosity 90 cSt or higher, the recommended test temperature is 82°C (179.6°F) instead of 54°C, and the volumes of oil, water, and emulsion are recorded after 60 minutes instead of 30 minutes.

FOAMING CHARACTERISTIC TEST (D 892)

Foaming characteristic is an important qualification test for hydraulic fluids, as many performance factors are dependent on this property. Many test procedures have been attempted [5–9] in the past to evaluate this property. Currently the ASTM test procedure is well accepted by industry. The test determines the foaming characteristic of hydraulic fluids at specified temperatures by empirically rating the foaming tendency and stability of the foam. The schematic diagram of the test apparatus is shown in fig. 3–4. The test is conducted in three

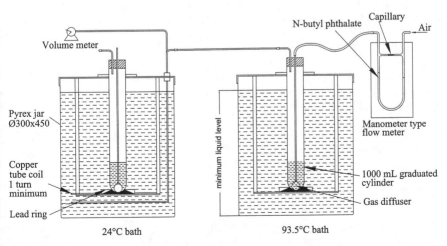

FIGURE 3–4 Foam Test Apparatus Setup

FIGURE 3–5 Dual Twin–Foam Test Baths Apparatus (Courtesy: Stanhope-Seta Limited, U.K.)

sequences. In sequence one, dry air is blown off at a constant rate (94 mL/min) through the fluid sample at 24°C (75°F) for five minutes. The sample is then allowed to settle for 10 minutes, and the volume of foam is measured at the end of the blowing and settling periods. In sequence two, the test is repeated on a second sample at 93.5°C (200°F). The volume of foam is measured at the end of a blowing period of five minutes and a settling period of 10 minutes. In the sequence three, after the test at 93.5°C, the foam is collapsed by stirring, the sample is cooled to 24°C (75°F), and the test is repeated as in sequence one. Again the foam volume is measured at the end of the blowing and settling periods.

The volume of air passed through in all the tests is 470 ± 25 mL. The readings taken in all three sequences are reported. The foam volume is normally around 10 to 30 mL at the end of blowing, depending on the viscosity of the oil. Premium grade hydraulic oils generally exhibit no foam after 10 minutes in all three sequences. In the test apparatus shown in fig. 3–5, all three sequences of the test can be conducted simultaneously.

AIR RELEASE TEST (D 3427)

The air release test is related to the antifoam property and is used to measure the ability of petroleum-based oils to separate entrained air. Though this test was originally intended for steam turbine oils, it is now widely accepted for hydraulic

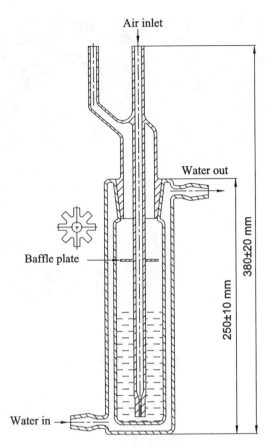

FIGURE 3–6 Air Release Test Apparatus

fluids. This test procedure came later to the antifoam test, and before it came under the ASTM label, other methods were also attempted [5–9]. The schematic diagram of the test apparatus is shown in fig. 3–6. The test is conducted at 25°C, 50°C, or 75°C, depending on the viscosity of the oil. For oil with a viscosity of less than nine cSt, 25°C is used; for a viscosity of 9–90 cSt, 50°C is used; and for a viscosity above 90 cSt, 75°C is used. In industrial hydraulics, 50°C (122°F) is the commonly used test temperature, as the viscosity of the oil used seldom exceeds the limit set for 50°C. The test consists of blowing compressed air (20 kpa) through 180 mL of test fluid at 50°C for seven minutes so as to form air-in-oil dispersion. After the airflow is stopped, the time required for the air entrained in the oil to reduce in volume to 0.2% is recorded as air separation time. Typical air separation time can vary between one and 15 minutes, depending upon the viscosity of the oil; the lower the viscosity, the shorter the separation time. The P.C.-based apparatus shown in fig. 3–7 can automatically record air separation time without any manual intervention.

A word of caution: The test involves only the release of air bubbles at atmospheric pressure, not air that is dissolved at high pressure and released when

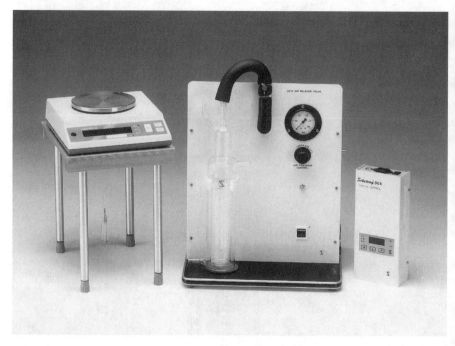

FIGURE 3–7 Air Release Value Apparatus (Courtesy: Stanhope-Seta Limited, U.K.)

the pressure is relaxed—a situation commonly encountered in high-pressure hydraulics.

ANTIWEAR PERFORMANCE TEST

Antiwear performance is an important property of hydraulic fluid, as the life of hydraulic components partially depends on this property. It is a test that has attracted many researchers [10–23]. The generally adopted procedure is the vane pump wear test. Kessler and Feldmann [23] have written a detailed account of contact conditions and types of loads and motions—in precise terms, what they call "tribological load," encountered in various hydraulic elements. Compared with other hydraulic elements, the contact conditions encountered in the vane pump are severe and hence taken as representative of a critical element in the hydraulic system (fig. 3–8). The line contact, high specific pressure, impact load, and high sliding action encountered between the vane and ring cause a situation that leads to severe wear, and so the vane pump is recognized as good ground to assess the antiwear performance of hydraulic fluid. This test is, however, time-consuming, elaborate, and expensive. Though attempts have been made to formulate alternative test methods, the Vickers V-104C/105C vane pump test is still the one, that enjoys widest industry acceptance. The fact that other national standards such as DIN and IP/BS also adopt this method lends credence to that.

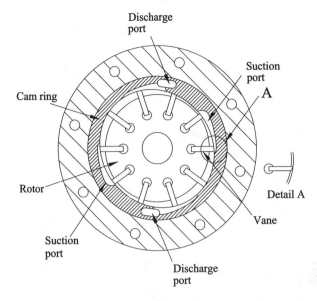

FIGURE 3–8   Vane Pump Showing Vane Contact Condition

Vane Pump Wear Tests

ASTM D 2882 Method The hydraulic circuit followed for the vane pump wear test per ASTM D 2882 is given in fig. 3–9 [1]. In this test, 11.4 liters (3 gal) of hydraulic oil is circulated through a Vickers V 104C or 105C vane pump for 100 hours at a pressure of 13.79 MPa (2,000 psi) and a 1,200 rpm speed, at a fluid inlet temperature of 65.6°C (150°F) for water-based fluids and at 40°C (104°F) for mineral oils and synthetic fluids if the viscosity is VG46 or less, otherwise at 79.4°C (175°F) for all other viscosities. At the end of the test, wear of ring and vanes in terms of weight loss is measured in mg. For most mineral hydraulic oils, the total weight loss does not exceed 50 mg. The quantity of test oil required is 18.9 liters (5 gallons).

DIN 51389/IP 281 Method The test conditions prescribed in the BS/IP and DIN differ slightly from the ASTM procedure, in that the test is conducted for 250 hours at 140 bar (2,030 psi) and 1,440 rpm [26, 27]. For water-based fluids, the test is conducted at 105 bar (1,522 psi), and the maximum temperature is limited to 60°C (140°F). A major difference between the two methods is that in the ASTM procedure, at the pump inlet the temperature of the fluid must be maintained at a specified level depending upon the viscosity of the test fluid. But in the DIN/IP method, for mineral oils the temperature of the fluid must be adjusted so that the viscosity of the oil at the pump inlet is 13 cSt. And for water-based fluids, it is 30 cSt for VG 32 to 68, and 60 cSt for VG 68 and above. This approach is better for assessing antiwear performance, as the viscosity of the oil and the consequent film thickness will not influence the wear process [19]. The additive used in the fluid will only predominantly influence wear, and this fits well with the main objective of this test—to assess the ability of the

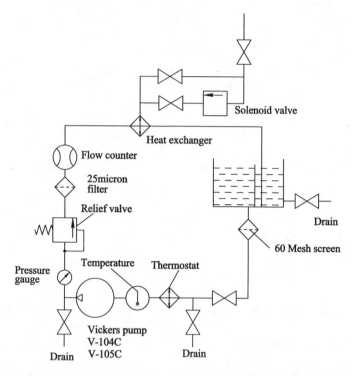

FIGURE 3–9 Schematic Diagram of Vane Pump Wear Test

antiwear additive present in the fluid. The weight loss observed may be high with this procedure; however, fluid rating is found to be in the same order with that of the ASTM test procedure.

Vickers 35 VQ 25A Vane Pump Test Vickers has also formulated one more vane pump test [16, 28] with a 35 VQ 25A vane pump, which is an accelerated test and more severe than the V104C pump test. The test is conducted at a pressure of 207 bar (3,000 psi) and a speed of 2,400 rpm at a temperature of 93°C (200°F) for 50 hours. The surface speed of the vane is more than twice that of the V104 vane pump test. For the fluid to qualify, the total weight loss of vanes is limited to 15 mg, and the weight loss of rings should be less than 75 mg. Apart from the usual weight loss criterion, the ring should not show any evidence of unusual wear or stress in the contact area, even though weight loss is lower than the prescribed limit. It should be noted that the assessment goes beyond weight loss, which many consider an inadequate criterion. Vickers recommends this test primarily for the qualification of antiwear fluids used in mobile applications. With the five fluids they have tested, Perez and Brenner have reported [15] that the ranking of fluids is the same in both V104C and 35VQ 25A vane pump tests, though weight loss is different.

Denison T6 Vane Pump Test Denison has also formulated a vane pump test (A-TP-30283) meant for severe applications to evaluate not only mineral oils but also water-based and biodegradable fluids [25]. The test is conducted at

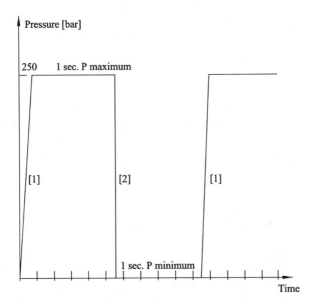

FIGURE 3–10   Pressure Cycle in Denison T-6 Vane Pump Test
1. Pressure rise = 3,500 < dp/dt < 7,500 bar/second
2. Pressure down = 15,000 < dp/dt < 30,000 bar/second

250 bar (3,625 psi), 1,700 rpm, and 80°C (176°F) for mineral oils; 45°C (113°F) for water-based fluids; and 70°C (158°F) for vegetable-and synthetic-based fluids. The test lasts a grueling 600 hours—300 hours without water and 300 hours with 1 percent distilled water added. Also, the pressure is not constant but cyclic—it switches between 10 bar and 250 bar every second (fig. 3–10). Evaluation is done by measuring a host of parameters: total weight loss of vanes; condition of the internal cam ring profile (seizing, burnishing, or ripples); condition of the port plate surface; viscosity; filterability; filter pressure drop; contamination level; and TAN value.

Though the V104C vane pump test is most widely used, the inadequacies of this test method are attracting the attention of researchers as well as standardization bodies [15,19]. Kunz and Broszeit have pointed out [19] that the V104C pump has served for a long time, but its obsolete design limits to rise the test pressure to, say, 210 bar, which is quite common in current industrial hydraulics. Another drawback they discuss is that the present test standards (ASTM and DIN/IP) do not specify the contamination level of the test fluid, and variation in this parameter can distort wear results in a set of tests, and therefore repeatability and reproducibility of wear results may suffer. They have suggested monitoring contamination levels during the test, as this is related to wear particles generated and also has an added advantage that the test need not be interrupted for any measurements. If a relationship between contamination level and wear limits could be established, the test procedure could be made simpler; this may also help reduce the length of the test. Of course, this calls for extensive experimental investigation and data generation and is a good case for further research.

In place of the V104C pump, use of the Vickers 20VQ5 pump is also being considered, as this pump can be operated at a higher pressure and speed [12,13,19]. The tests conducted by Perez and Brenner [15] with various vane pumps under different test conditions indicate there is a correlation between the various test methods and the fluids they have tested, i.e., the ranking of the fluids is the same in all the tests.

Kessler and Feldmann have come up with a novel test concept that moves away from conventional pump tests [23]. Their concept, called the flywheel test, simulates the contact conditions of both the vane tip ring and the piston-cylinder—it simulates both vane pump and piston pump operation, and the load and speed can be varied independently. Initial test results are encouraging, and further work may provide a workable test.

FZG Gear Rig Wear Test (DIN 51354-2)

In addition to these pump tests, the FZG gear test formulated by DIN is also gaining acceptance as an important antiwear performance test. It is an additional test for evaluating antiwear performance, formulated according to German standards; a technically equivalent IP standard also exists [29, 30].

Basically, it is a test to evaluate the load-carrying ability of the antiwear additive used in any lubricating oil, not only hydraulic oils, and is considered a more severe test than the vane pump wear test. The basic principle of this test is that the load-carrying capacity of the lubricant is defined by the maximum load, which can be sustained by the lubricant without failure of gear tooth surface. So if the load-carrying capacity of the lubricant is exceeded, it fails to protect the gears and the surface of the teeth is damaged. It is a parameter featured in the DIN HLP grade hydraulic oil specifications and is more commonly used in continental Europe. The test is conducted in an FZG gear-testing machine (fig. 3–11), and for obvious reasons the results obtained from this test are different from those of the vane pump test. Though it does not simulate the actual working condition of hydraulic pumps, there is a belief that this test relates to the behavior of piston pumps.

This test is described as A/ 8.3/ 90: A refers to a particular tooth profile, 8.3 to the sliding speed at pitch circle in m/sec, and 90 to the initial oil temperature in °C from load stage four. A pair of specially designed case-hardened gears is run at the test speed and temperature in a dip lubrication system. The test prescribes 12 load stages. At every load stage, the torque on the pinion and tooth normal force is progressively increased. The test is carried out by first running the gears in the test oil for 15 minutes applying the load pertaining to stage one as prescribed in the standard. The test is repeated till stage four. At the end of stage four, the pinion is inspected for damage without removing the gears, and the tooth conditions are recorded. The test is continued in incremental load stages, and the gears are inspected at the end of every stage. The test is continued until the failure load stage is reached. If no damage occurs by the end of load stage 12, the test is terminated. The failure stage is indicated when the estimated sum of total width of the damaged areas on all the gear teeth faces equals or exceeds one gear tooth width (20 mm). The results are expressed in terms of maximum

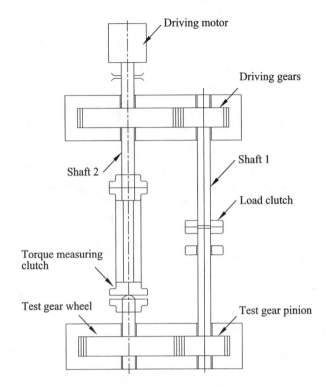

FIGURE 3–11 Schematic Diagram of FZG Gear Test Rig

load applied on the gears (number of stages). Most premium grade antiwear hydraulic oils pass 10 stages. Many leading hydraulic oil manufacturers now commonly specify the FZG gear rig wear test as a parameter in the specification. Hydraulic oils passing stage 12 in this test are available.

As mentioned earlier, formulation of test standards is a dynamic process; the V104/105 vane pump test, which has been in existence for three decades, is under scrutiny and alternative methods are being worked out. Formulating a reliable and acceptable test procedure is a difficult task that involves many organizations, cuts across continents, and generates a great deal of data. It is a key development stage of any technology. In the coming years a new standard prescribing a single test may emerge to assess the antiwear performance of hydraulic fluids, a test that takes into account all possible operating and field conditions, which would have worldwide industrial acceptance.

OXIDATION TEST (D 943)

The oxidation test is one of the key tests used by lubricant manufacturers during the development stage. This test was used to evaluate the effectiveness of oxidation inhibitors used in turbine oil, called TOST (turbine oxidation stability test). It is now widely used to predict the oxidation life of both hydraulic and turbine oils. The oil sample is subjected to a temperature of 95°C (203°F) in the presence of water, oxygen, and an iron-copper catalyst.

The test consists of passing the oxygen at a rate of three liters/hour into a mixture of 300 mL test oil and 60 mL distilled water, in the presence of an electrolytic copper wire that is 99.9% pure and has a diameter of 1.69 mm (0.064″), and a low-metalloid steel wire that has a diameter of 1.59 mm (0.0625″) as a catalyst at 95°C (203°F) in the oxidation cell. The details of the oxidation cell are given in the standard. At the end of 1,000 hours, the total acid number (TAN), sludge, and copper and iron mass loss are reported. At the end of 1,000 hours, TAN value generally does not exceed one. The test is also continued until TAN value reaches two, and the hours noted are the oxidation life of the oil. Oils with a life of 2,000 hours and above are more common today. In fact, the so-called new generation of zinc-free oils have a life of 3,000 hours—this is indeed a remarkable performance achievement.

Because biodegradable hydraulic fluids like triglycerides and synthetic esters (Chapter 7) are sensitive to water and prone to hydrolysis, this test procedure is not suitable, as it is conducted in the presence of water. For these fluids, the Baader oxidation test as prescribed in DIN 51554 is used to assess the oxidation stability or aging characteristics [31]. In this procedure, the fluid is subjected to a temperature of 95°C (203°F) in the presence of air and copper in the Baader test apparatus for 72 hours. At the end of 72 hours, the increase in viscosity of the fluid is measured; this should not be more than 20% of the original value.

NEUTRALIZATION NUMBER (NN) BY COLOR-INDICATOR TITRATION (D 974)

This method is used to measure the total acid number to indicate relative changes that occur in oil during use under oxidizing conditions. Excessively dark oil cannot be analyzed by this test method. As explained earlier, this test is extensively used to analyze oils-in-service as a part of maintenance programs, and is also used as a guide in quality control of lubricating oil formulations.

The sample oil is dissolved in a mixture of toluene and isopropyl alcohol that contains a small amount of water. This single-phase solution is titrated at room temperature, with standard alcoholic acid solution to the end point indicated by the color change of the added p-naphtholbenzein solution. To determine the total acid number, this solution is titrated with potassium hydroxide solution, using methyl orange as an indicator. The end point of titration is indicated when the solution changes from orange to green or green to brown. Total acid number is calculated by the procedure prescribed in the standard.

Total acid number is the quantity of base expressed in milligrams of potassium hydroxide (KOH) that is required to titrate all acidic constituents present in one gram of sample oil (mg KOH/g). TAN is often expressed simply as a number.

HYDROLYTIC STABILITY TEST (D 2619)

Hydrolytic stability is the measure of a hydraulic oil's ability to resist degradation when contaminated with moisture or water. This test is used for petroleum- or synthetic-based hydraulic fluids. In this test, 75 g of test fluid along with 25 g of distilled water and a test copper specimen of grade (QQ-C 576A), 16-22 B&S gage and size 13 × 51 mm are sealed in a 200 mL beverage bottle. The bottle is rotated end to end at five rpm for 48 hours in an oven at 93°C (200°F). At the end of the test the weight change in copper specimen in mg/cm², percentage change in viscosity at 40°C, and acid number of fluid and acidity of water layer are measured. Safety measures outlined in the standard must be scrupulously followed. Typical values for premium grade oils are given in Table 2–3. Water-based fluids can also be evaluated by this method without adding additional water to 100 g of sample fluid.

WATER CONTENT BY DISTILLATION (D 95)

This test is also known as the Dean and Stark method. Water content in the hydraulic oil can be measured accurately in the Dean and Stark apparatus shown in fig. 3–12. A measured quantity of oil is dissolved in a water-immiscible hydrocarbon solvent such as xylene and heated in a distillation flask. The boiling solvent drives the water vapor along with its own vapor into a condenser, where the vapors condense and fall into a calibrated trap. The water, being heavy, settles at the bottom (this section of the trap is graduated), where its volume can be measured directly, while the solvent overflows back to the distillation flask. Water present is expressed as volume or weight percent according to the basis on which the sample is taken. The test can be used for both virgin and used oils. The limit for water content is 0.1%.

FILTERABILITY TEST

There are no ASTM or other national standards like DIN or BS for evaluating the filterability of hydraulic oil. For a long time the Denison filterability test procedure TP 02100-A was widely used to evaluate filter plugging [32, 33]. Today AFNOR (a French standard) and ISO filterability test standards are also used in many specifications [34, 35]. The Denison procedure is a simpler method compared with the other two. All three test procedures will be discussed here.

Denison Method

The test apparatus used in the Denison method is given in fig. 3–13. The test involves the filtration of a 75 mL sample (from a 100 mL sample) of dry oil as well as a 2 percent (by volume) water sample mixed with oil under a vacuum of 65 cm (26″) Hg/1.3 psi through a 1.2 μm membrane filter (9.6 cm² area),

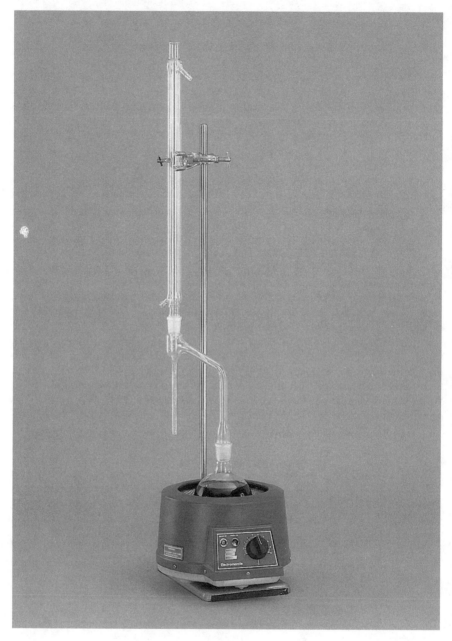

FIGURE 3–12 Dean and Stark Apparatus for Measuring Water Content
(Courtesy: Stanhope-Seta Limited, U.K.)

100 mL sample oil

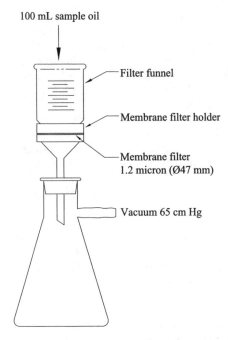

Filter funnel

Membrane filter holder

Membrane filter
1.2 micron (Ø47 mm)

Vacuum 65 cm Hg

FIGURE 3–13 Denison Filterability Test Apparatus

at a temperature between 18°C and 24°C (65°F to 75°F). The time required to filter 75 mL of sample fluid is recorded. The test is discontinued if the time for filtration exceeds 600 seconds, and filtration time is recorded as 600+. The fluid is considered to pass if the time taken for the wet sample (with water) is less than twice the time taken for the dry sample. The wet sample test is significant, as a small amount of water leaking into the system through surrounding air or hoses can never be ruled out. The filtration time varies depending upon the viscosity of the oil—low viscosity oil takes less time than high viscosity oil. Many leading hydraulic fluid manufacturers make it a point to qualify their oil for this test. Typical values for this test are given in Table 2–3.

AFNOR Method

The AFNOR test is also conducted in two parts, with and without added water. In the first part, 320 mL of test fluid is placed inside a container, which is pressurized by air to 760 mm (30″) Hg. The fluid is filtered through a 0.8 mm membrane filter and collected in a 500 mL graduated cylinder. The time taken for filtering 300 mL fluid is noted (T_{300}). The test is repeated for filtering 200 mL, 100 mL, and 50 mL of fluid, and the corresponding time taken for filtering is noted. The filterability index (FI) is calculated by using the following formula:

$$FI = (T_{300} - T_{200})/2(T_{100} - T_{50})$$

In the second part of the test, 60 mL of distilled water is added to the test fluid. The mixture is stirred for five minutes at room temperature in a screw-top flask and placed inside an oven for 72 hours at 70°C (158°F). After 72 hours, the jar is cooled at room temperature (the typical room temperature is 22°C/70°F) for 24 hours. The fluid is agitated, and the test is repeated as in part one. If the filterability index is one, the fluid is considered best; if the index is two or less, the fluid is rated as acceptable.

The basic difference between these two procedures is that in the Denison method the flow is effected by vacuum, and in the AFNOR the flow is effected by mild pressure. Not much work seems to have been done comparing the test methods, but a correlation between the two methods does not seem to exist [46].

ISO 13357-2 Method

The recently released ISO test method [35] is similar to the AFNOR procedure in that both methods use pressure for the flow of oil, and filtration time is noted at the end of series of volumes. The current ISO document is limited to dry oils only, and the test procedure for wet oils is under way. The test is conducted in two stages. Stage one filterability is expressed as a percentage ratio between 240 mL and the volume of oil actually filtered in the time that 240 mL would have theoretically taken, assuming no plugging of the membrane. Stage two filterability is expressed as the percentage ratio between the flow rate near the start of filtration and the flow rate between 200 mL of filtered volume. The second stage is a severe test in that it is sensitive to the presence of gels or silts in the oil, which are generally formed as oil ages, if not present in virgin oil. The oil that passes through the stage two test is likely to perform well without any problems even under conditions of fine filtration (less than 5 μm).

For details of the test apparatus, readers may refer to the standard. The test is conducted by filtering the test oil through a membrane filter 47 mm diameter with a mean pore size of 0.8 μm at a pressure of 50 kPa/7.25 psi (for VG12 and less), or 100 kPa/14.5 psi (for VG 32 and VG 46) or 200 kPa/29 psi (for VG 68 and 100). Using the dual-stop facility of the timing device, the time taken to filter 10 mL (T_{10}), 50mL (T_{50}), 200 mL (T_{200}), and 300 mL (T_{300}) of oil are noted. After the T_{50} value is available, the time T_v is calculated per the relation given below, and the filtered volume is recorded for the calculated time T_v. Then the filterability is calculated:

Stage I filterability

$$F_1 = [(V - 10)/240]\,100$$

Where V = volume collected at T_v in milliliters

$$T_v = 6(T_{50} - T_{10}) + T_{10} \text{ in seconds}$$

Stage II filterability

$$F_{II} = [2.5(T_{50} - T_{10})/(T_{300} - T_{200})]\,100$$

If the filtration time exceeds 7,200 seconds (two hours), the test should be discontinued and reported "unfilterable".

SEAL COMPATIBILITY TEST (ISO 6072)

A reliable evaluation method for this property was elusive for a long time. The British Fluid Power Association (then AHEM) initiated a procedure to test the compatibility of seals in the 1970s, and now the revised and updated version is available as ISO: 6072 [36, 37]. It is also a useful test for manufacturers to evaluate additives and fluid formulation with regard to the compatibility of seal materials. The test method given evaluates the effect of both aqueous- as well as nonaqueous-based hydraulic fluids on the test elastomeric materials under controlled test conditions. The changes in volume, hardness, tensile strength, and elongation at break are measured before and after the immersion of the test elastomer in a test fluid under specified test conditions to establish an elastomer compatibility index (ECI).

The standard prescribes three types of test elastomers for evaluation: acrylonitrile-butadiene rubber (NBR 1); fluoroelastomer (FPM 1); and ethylene propylene diene rubber (EPDM 1). The test elastomer is chosen depending upon the test fluid. For example, NBR 1 or FPM 1 is chosen for mineral oils and water-based fluids, while EPDM 1 or FPM 1 is used for phosphate esters. A rectangular test piece that is 50 × 25 mm (~ 2″ × 1″) and two mm thick, or a circular one with a 36 mm diameter that is two mm thick can be used. The test piece is kept totally immersed in the test fluid in a glass jar for 168 hours at a temperature of 100°C (212°F) for nonaqueous fluids, and at a temperature of 60°C (140°F) for aqueous fluids. At the end of the immersion period, the percentage change in volume, tensile strength, elongation, and change in hardness are measured and calculated per the procedure given in the standard. Based on this data, ECI is expressed as given in the standard.

The standard also prescribes a simpler and quicker method to be used only for mineral oils. The test piece used is a standard ring with a 25 mm internal diameter and a square cross section of 1.5 mm. The test piece is immersed in 50 to 55 mL of the test oil for 22 hours at 100°C (212°F), and the volume change is calculated before and after the immersion by measuring the change in internal diameter before and after the immersion by means of a tapered rod gauge. The procedure for calculating the volume change index (VCI) is given in the standard. Though the percentage change in the volume of the test elastomer may differ in both methods, ranking the results by either method is the same. By this method, an index value below six is an indication that the seal is susceptible to shrinkage and hardening; an index of more than 30 is an indication that the seal can soften and swell.

The standard test method described above meets most industrial applications; however, it is not an end itself. It may not always adequately address all requirements of critical applications. Murray, for example, reports various test methods used for assessing the compatibility of seals for defense applications [39]. Hence users can always modify the test conditions and methods to suit specific applications. In fact, this is the way new test methods evolve and replace the old ones. And this also leads to performance improvements of the products.

THERMAL STABILITY TEST

The test procedure for this property came into existence at the initiation of Cincinnati Machine, a leading machine tool manufacturer in the U.S. [42]. ASTM then adopted one of the procedures. There are two test procedures prescribed by ASTM: D 2160 [41] is used to determine the thermal stability of hydraulic fluids; D 2070 [40] is used to evaluate mineral hydraulic oils. There are no other national standards for evaluating thermal stability.

ASTM D 2160 Method

The test procedure prescribed in D 2160 consciously precludes the catalytic effects of air and contaminants such as water and metals, and takes into account only the effects of temperature. The test involves heating 20 mL of sample oil in a glass tube at a temperature of 316°C (600°F) under a pressure of 0.13 kPa (a mild vacuum is created to preclude air) for six hours. Then, at ambient temperature, the visual appearance of the test cell, change in TAN and percentage change in viscosity in cSt at 40°C (104°F) are reported.

ASTM D 2070 Method

Unlike the D 2160 procedure, the D 2070 test evaluates the oil in the presence of catalyst, copper, and steel at 135°C (270°F). Electrolytic copper that is 99.9 percent pure, with a 6.35 mm diameter ×76 mm length (0.25' × 3") and a 1 % carbon steel (AISI W-1) rod immersed in the 200 mL of sample oil kept in a 250 mL Griffin beaker is placed inside an aluminum block in an electric gravity convection oven for 168 hours at 135°C. At the completion of test, the sample is allowed to cool to room temperature, and the color of the copper and steel rods, and the total sludge in mg/100 mL oil, is reported. This test procedure is adopted from the Cincinnati Machine's test procedure.

Cincinnati Machine's Method

The test procedure of Cincinnati Machine is more widely quoted, and it is almost identical to the ASTM D 2070 test, except that more parameters are reported [42]. In industry circles it is viewed as a very stringent test. It is also a way to evaluate the multimetal compatibility of the fluid.

Cincinnati Machine's test [42] has two parts: test procedure A and test procedure B. Procedure A is used for the evaluation of Rust&Oxidation (R&O) inhibited oil and antiwear oil. In procedure A, cleaned and pre-weighed (nearest to 0.1 mg) electrolytic copper that is 99.9 percent pure, and steel rods (AISI

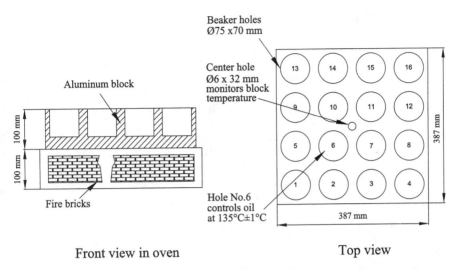

Front view in oven

Top view

FIGURE 3–14 Aluminum Test Fixture (Cincinnati Machine's Test)

W-1, 1 percent carbon) that are 0.25″ diameter and 3″ long are placed in a 250 mL Griffin beaker which contains 200 mL of sample oil. The beaker and the contents are placed in an aluminum block (fig. 3–14) in an electric gravity convection oven for 168 hours at a temperature of 135°C (270°F). The aluminum fixture helps keep the temperature distribution uniform. At the end of 168 hours, the samples are allowed to cool at room temperature and evaluated. The results are expressed in terms of increase in viscosity, as well as TAN, the appearance of copper and steel specimens, weight loss of the copper and steel rods, formation of sludge and deposits on steel and copper, and weight of total sludge/100 mL. Typical limits for this test are given in Table 2–3.

Procedure B, is similar to A, except that 80 mL of oil used, and it is baked at 101°C (213.8°F) for 24 hours. The results are expressed in the same way as those procedure A. Procedure B is less severe and is used only for evaluation of slide way cum hydraulic oil. Readers may note that temperature and duration of the test are reduced for this oil because polar additives (fatty acids) present in the oil cannot withstand the severe test conditions used for HM type oils.

SHEAR STABILITY TEST

As described in the previous chapter, oil fortified with VI improvers (long-chain polymers) exhibit temporary as well as permanent loss of viscosity due to lack of shear stability. Testing evaluates the shear stability of oils containing VI improvers. It is used for evaluating both crankcase oils and HVI hydraulic oils, as both these oils contain the additive. The more commonly used test procedure in Europe and the U.K. is IP 294 [43]. The IP test method is technically equivalent to DIN 51382.

IP 294 Method

In this test, the high shear condition encountered in the actual working of the oil is simulated by mechanically stressing the oil, i.e., passing the oil through a diesel injection fuel pump and a diesel injector nozzle. The oil is passed through the prescribed test circuit for predetermined passes, which is normally 250 passes for the hydraulic oil (it can be as high as 500 passes), and only 30 passes for crankcase oils. The number of passes is high for hydraulic oils because they are more shear stable than crankcase oils. The oil is passed through the injector nozzle at a pressure of 175 bar (2,540 psi), maintaining the flow rate and temperature at 170 mL/min and 30°C to 35°C (86°F to 95°F), respectively. After the completion of the predetermined passes, the sample oil is removed, and the kinematic viscosity at 40°C and 100°C are determined. These viscosity values are used to calculate, on a percentage basis, the extent of oil degradation due to shear forces in the test. The numbers of passes, VI before and after testing, and the percentage decrease in kinematic viscosity at temperature are reported.

One could enlarge the scope of the standard test by making suitable modifications in the test procedure to suit any particular application. An interesting example is the modified test procedure of IP 294 used by the U.K.'s Ministry of Defense to assess the shear stability of polyglycol [47]. After 250 passes, the water content in the fluid is readjusted to its original value, and then the kinematic viscosity of the fluid is measured at 40°C. The percentage change in kinematic viscosity at 40°C before and after 250 passes is reported, and the percentage change in water content after shearing is also reported. For the fluid to pass this test, the decrease in viscosity should not exceed 3 cSt.

ASTM D 3945 Method

ASTM D 3945 [44] prescribes two test procedures: procedure A, similar to that of IP/DIN method (the same test equipment is used); and procedure B, where the test equipment used (fuel injector shear stability test—FISST—equipment) and the test conditions are different. It is a more severe test than procedure A; for example, the test pressure is 20.7 MPa (3,000 psi), and the delivery rate is 534 mL/minute, but the number of cycles are only 20. However, evaluation by both the methods ranks the fluid in the same order, and both methods show comparable repeatability and reproducibility.

There is also another ASTM test procedure per D 5621 [45] specifically for hydraulic fluids. In this method, shear condition of the fluid is simulated by subjecting the fluid to sonic oscillations. Thirty mL of sample oil contained in a 50 mL Griffin beaker is irradiated in a sonic oscillator at a preselected power setting for 40 minutes at 0°C. After irradiation, the viscosity of the sample oil at 40°C is measured. Viscosity of the initial and the irradiated oil at 40°C is reported. Correlation between the test and the actual hydraulic applications has been established.

CONCLUSION

The test procedures described here give a broader, in-depth view of the technology of making high-performance hydraulic fluids, and also a better understanding of the various properties of hydraulic fluids. The number of tests the fluids must go through to qualify for various performance characteristics are indeed mind-boggling. Remarkably, technology development has been able to cope with the changing performance criteria set by industry. The combined efforts of standardization bodies, researchers, and lubricant manufacturers have made this happen.

Chapter 4

HIGH WATER CONTENT FLUIDS

High water content fluids (HWCF) as a class separate from fire resistant fluids came into prominence in the 1970s, mainly as a replacement to mineral oils. The oil crisis of the early 1970s fueled the R& D work on HWCF with a view to conserving petroleum products. Automotive industries in the U.S. [1–3] looked at this crisis as an opportunity to find a replacement to mineral oil with a less expensive and easily available fluid, as U.S. automotive industry's consumption was as high as 3.5 million liters per annum. Automotive industries, machine tool builders, hydraulic fluid and hydraulic hardware manufactures charted a development program to make HWCF a viable alternative fluid to mineral oil. The initiative taken at that time did bear fruit, and today HWCF is here to stay, though its application is not as wide as was envisaged.

TYPES OF HWCF

Per ISO classification (Appendix Table A-1), HWCF is categorized as a fire resistant fluid, though, as mentioned earlier, HWCF entered as a class by itself and is under the category of HFA fire resistant fluids. Any fluid that contains more than 80 percent water is classified as HWCF. However, hydraulic systems using HWCF normally contain 95 percent water. Broadly there are two types of fluids under HFA. They are:

1. **Oil–in–Water Emulsion:** This type is categorized as an HFA-E fluid. It is a mixture of water and oil. This is similar to milky emulsion or soluble oil used in metal cutting operations, such as turning or grinding. Oil droplets of around 50 μm are dispersed in a continuous medium of water and give a milky appearance. Unsatisfactory performance of this fluid led to the development of a new type of emulsion known as "micro emulsion." Soluble oil is now seldom used, and micro emulsion is taking its place.

    Micro emulsion is the same as soluble oil, except that the oil droplet size is much smaller, i.e., around 2 μm, and the emulsion is translucent. It has less tendency for phase separation (oil and water) than traditional soluble oil and exhibits better emulsion stability. It also provides better lubrication and antiwear performance. Micro emulsions are now available with synthetic esters as concentrate (instead of mineral oil) to meet ecological standards [5].

Another fluid available in this category is thickened emulsion [6,7], which is available in viscosity ranging from 10 to 46 cSt. Thickened fluids are suitable even for servo applications. Water content in this fluid is less than in micro emulsion, ranging from 70 percent to 80 percent in order to keep the viscosity level high.

2. **Synthetic Solution:** This is categorized as an HFA-S fluid, and it is a nonpetroleum product. It is a mixture of water and synthetic oil that forms a solution. In essence, it is a host of chemicals dissolved in water, and the particle sizes are generally submicron. Being true solutions, they are transparent. Water content in this fluid is generally around 95 percent. One advantage of this fluid is that it is more tolerant to hard water. In addition, the solution is free from the problems of oil separation and bacterial growth.

ADDITIVES FOR ENHANCING PERFORMANCE CHARACTERISTICS

As in mineral oils, additives are blended to enhance the performance of HWCF. Some of the additives used in HWCF only counter the effects of high water content of the fluid. Details on some of the prominent additives used in HWCF are given below.

- **Emulsifier:** This is used to emulsify the oil with water. Emulsifier breaks the oil into small droplets and disperses it in the water. Emulsifier also keeps the oil droplets suspended in the water. In simple terms, it helps the oil mix with water.
- **Biocides:** High water content in the fluid provides a congenial atmosphere for all sorts of microorganism to breed, grow, and thrive. Biocides are added to check this undesirable property.
- **Corrosion and Rust Inhibitors:** Mineral oils inherently provide antirust characteristics, while HWCF, because of its high water content, inherently promotes corrosion and rust formation. Corrosion inhibitors are added to protect from corrosion in the liquid as well as the vapor phase.
- **Polar Compounds:** Owing to the poor lubricating property of HWCF, polar compounds (fatty materials) are added to enhance the lubricating property of the fluid.
- **Thickeners:** These additives are used in so-called thickened emulsion. To mitigate the effects arising from low viscosity, thickeners are added so as to raise the viscosity level to that of mineral oils. Conventional thickeners, such as long-chain polymers, exhibit poor shear stability under high shear conditions; for example, as in high-speed pumps or throttle valves—hence thickened fluids initially were not successful.

But the new generation of thickeners that form a mechanical matrix in the fluid exhibit higher stability.
- **Antiwear and Antifoam Additives:** They perform the same functions as in mineral oil.

These additives and the intensive development work on HWCF have made HWCF a viable hydraulic fluid. However, as the fluid is drastically different from mineral oil, a thorough understanding of HWCF technology is essential to realize the full benefits of HWCF. HWCF's inherent drawbacks, and maintenance of these fluids [6–17], are outlined here so that users can devise a workable plan of action to ensure the successful introduction and application of HWCF.

BASIC DRAWBACKS

Low Viscosity and High Specific Gravity

Low viscosity of HWCF results in high leakage and inadequate fluid film formations, which cause higher wear to pumps and valve components, thus limiting the operating pressure and speed of pumps. The recommended maximum pressure and speed for pumps is around 100 bar (1,450 psi) and 1,500 rpm (limited by the allowable sliding velocity of the vane tip). However, piston pumps operating at higher pressure, as high as 500 bar (7,250 psi) are in use [18]. These pumps are generally of special design, or they are operated at derated life. Low viscosity of the fluid also adversely affects the fine control of motion through flow control valves unless the valves are designed with much finer clearance—up to four times finer than normal valves. In many applications, poppet valves are found to be a better choice than spool type valves.

Low viscosity coupled with high specific gravity increases the Reynolds number 12 times that of mineral oil, and this results in turbulent flow rather than laminar flow. The high velocity turbulent flow can cause erosion to the metal surfaces.

The specific gravity of HWCF is slightly more than one, compared with 0.88 for mineral oil; this affects the suction characteristics, and the pump has to do more work for a given delivery. Suction filters are not generally recommended, as they would consume much of the available vacuum in the suction side and create cavitation problems. However, a coarse suction strainer of rating around 150 μm with a bigger area can be used. Higher specific gravity of the fluid also keeps many particles in suspension. Incorporating a return line filter of finer rating is essential to lessen this problem.

High Vapor Pressure and Cavitation

When the evaporization of fluid takes place, vapor pressure is the pressure of vapor in equilibrium above the fluid at a given pressure and temperature. Unlike mineral oil, HWCF has a high vapor pressure, and as such even a low-level depressurization in the suction side can bring the fluid to a vapor state. As a result, cavities filled with vapor will be formed in the liquid in the suction side, and when they move to high-pressure areas inside the pump, cavities will collapse, generating powerful shock waves. This phenomenon is known as cavitation. These shock waves can cause craters even in hardened steel.

Pressure loss occurs in the suction inlet, strainer, and suction pipes; an even higher pressure loss occurs inside the pump. When all these losses are added up, the pressure in the suction side can go down to as low as 0.2 bar (2.9 psi). At this pressure, the boiling point of the water will come down to around 50°C (122°F)—half that of the regular boiling point. Boiling and sudden vaporization of water can take place, which can lead to severe cavitation. The surface affected by cavitation is more susceptible to corrosion, and hence the combined effects of cavitation and corrosion can cause significant damage to the pump.

Flooding the suction inlet with pressurized fluid by means of an overhead tank or through a booster pump can alleviate this problem. Maintaining a pressure of around 0.5 bar (7.25 psi) at the suction side through a booster pump can significantly enhance the performance of the pump, although whether a booster pump is necessary is controversial. Of course, caution should be exercised that the booster pump does not suffer from cavitation!

Corrosion

The fluid in the liquid phase generally does not cause corrosion. So long as the fluid is maintained well and the components are kept in contact with the fluid, rust-inhibited fluid gives adequate corrosion protection. But the fluid in the vapor phase tends to cause corrosion despite the presence of vapor phase rust inhibitors. Vapor phase rust inhibitors evaporate away when the fluid is in vapor phase, so the fluid is slowly depleted of rust inhibitor. As this process continues, the fluid may ultimately be left without any vapor phase rust inhibitor, thus making the fluid corrosive.

A recommended solution is to operate the system with a closed tank so that evaporation can be effectively prevented; this may also reduce the rate of evaporation of the fluid. Another method to prevent corrosion is to paint the tank. HWCF acts as a solvent on most paints, but epoxy paint with compatible primer is effective against corrosion. Alternatively, a stainless steel tank, although expensive, is impervious to corrosion.

Filterability

The method of sizing and the filter rating used for mineral oils can also be used for HWCF. But a slightly bigger filter is desirable, to reduce the frequency of filter changing. A pressure filter of 10 μm rating is the usual recommendation, though a filter rated at 3 μm is also effective with HWCF. Most filter manufacturers offer filters compatible with HWCF. Incorporation of return line filters of around a

25 μm rating is another recommended practice for HWCF, as accumulations of contaminants are more severe with HWCF than with mineral oils.

Seal Compatibility

The elastomers used with mineral oils are generally compatible with HWCF. But seals made from materials like cork, leather, and asbestos are not suitable because they absorb water. There are different grades of elastomers in the same class; some are suitable but others are not. For example, high nitrile is compatible, but low nitrile is not; this is also true with many grades of urethane. Generally, most fluorinated elastomers, such as viton and neoprene, are suitable. Seal compatibility differs very much in the fluids of one supplier to the next, as the chemistry of the fluids can be different. Usually, the information given by the fluid manufacturers contain a list of seal materials that are compatible with their fluids. As these lists are based on their own test results, their recommendations are quite reliable.

Metal Compatibility

Materials such as magnesium, cadmium, zinc, lead, and aluminum cannot be used because of the alkaline nature of HWCF. Copper should be avoided as well, as it can become a source of galvanic action. Also, copper wear particles can destroy emulsifiers. Here again, the fluid manufacturer's recommendations can provide a sound basis for HWCF system design.

The problems associated with water and the compatible hardware requirements have been overcome to a large extent through technology. But the problems associated with maintenance, which pertain mainly to human error have slowed the growth of HWCF applications. The issues connected with maintenance are outlined in the succeeding paragraphs.

MAINTENANCE ISSUES

Water Quality

Water used in HWCF systems should be soft; water hardness of 250 ppm and above is not recommended. Hard water will interfere with emulsification and also will make the fluid unstable. This problem is more pronounced with emulsion type fluids. Even in synthetic solutions, calcium and magnesium salts present in hard water can interact with the additives and form insoluble substances, causing filter plugging. But the synthetic solutions marketed today are more tolerant of hard water. Deionized water can often solve this problem, but for some fluids it may be too soft. Some fluids require a minimum hardness of, say, 50 ppm to function properly. In such cases, fluid suppliers should be consulted and their recommendation followed.

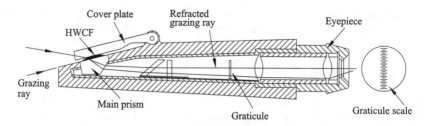

Cover plate   Refracted
               grazing ray   Eyepiece

HWCF

Grazing
ray

Main prism

Graticule

Graticule scale

FIGURE 4–1   Pocket Refractometer

Even if water quality is good, proper mixing of the fluid with water is not a simple matter. Water should never be added to the concentrate; if it is, the resulting emulsion would invert. The concentrate is slowly added to the water with continuous stirring. In fact, the fluid stability (phase separation) problem can be considerably lessened by proper mixing. To avoid human lapses, mechanizing the mixing with a pump and mixing valve may be desirable.

Even if proper water quality and mixing are maintained, the problem does not end there. To counter the effect of high evaporation of HWCF, the periodic addition of makeup water can reintroduce the problem. It is advisable to use only distilled or deionized water for makeup fluid; tap water is generally hard.

Maintaining Fluid Quality

Maintaining the fluid concentration at a level of a 95/5 ratio, or the ratio recommended by the manufacturer, is a prime step in the maintenance of the fluid. Increase in the water content may adversely affect the lubricating property of the fluid, reducing pump life. On the other hand, increased oil content will not bring any added benefit, and it may increase the fluid cost. The concentration level can be checked with a pocket refractometer (fig. 4–1). The concentration level should be restored to the original level by adding premixed solution of the right concentration. Adding water or oil directly to the fluid can result in improper mixing.

pH Level

The pH value of a fluid is a measure of both the alkalinity and acidity level of the fluid. For example, the pH of water of is seven—it indicates that the fluid is neutral, i.e., the fluid is neither acidic nor alkaline. A pH value below seven indicates that the fluid is acidic, and above seven indicates alkalinity. The pH of HWCF is normally maintained between 8.5 and 10. A fall in the pH value is an indication that the fluid is becoming acidic, which means the fluid is unstable or is contaminated with bacteria.

A pH meter can be used to measure the pH. Any fall in the pH value can be restored to the original level by adding an alkaline additive. However, it is important to understand the causes of this problem and to take remedial measures if necessary.

Biodegradation Control

The emulsifier and contaminants present in HWCF make the fluid vulnerable to bacterial attack. Though biocides are added to the fluid to control this, for any given operating condition it is difficult to tell how far and how long the additive will be effective. Periodic checking of bacterial growth by means of dipslides, also called as bio sticks, is essential [19].

Dipslides are dipped into the fluid, and after excess fluid is removed, they are kept inside an airtight tube and incubated for 24 hours at a temperature between 25°C and 35°C (77°F and 95°F). The infected dipslide is then compared with a model chart provided by the manufacturer, which gives an approximate count of bacteria per mL of fluid. Dipslides are available to identify both aerobic bacteria (those that thrive in the presence of oxygen) as well as anaerobic bacteria (those that thrive without oxygen). Periodic testing will indicate the pattern of growth, bacteria, and suitable remedial measures must be taken—adding the right type and quantity of biocides. This test will also indicate whether the fluid has reached ultimate degradation and must be discarded. An unpleasant odor and a lower pH value will generally accompany bacterial contamination in excess of more than 10^4. When the bacterial level reaches 10^6, the fluid must be discarded.

Good leakage control, dedicated housekeeping, and periodic filter servicing are also essential parts of maintenance, which also add to the burden and increased cost of maintenance.

CHANGEOVER PROCEDURE FROM MINERAL OIL TO HWCF

The successful application of HWCF requires a well-designed changeover procedure [10–13]. The procedure discussed below will help readers understand the implications of HWCF and gear-up in a better way. The changeover procedure has three stages: pre-changeover, changeover, and post-changeover.

Pre-changeover

Operating personnel accustomed to mineral oils may experience culture shock when HWCF is introduced if they are not educated and trained well, as HWCF is distinctively different from mineral oil. Hence, as a first step, a well-structured training program should be conducted for all appropriate personnel from design, plant engineering, laboratories, purchase, etc., as well as the fluid supplier. Once all the concerned people are conversant with HWCF, the checklist given below should be followed.

Checklist

1. *Reservoir:* Check whether reservoir size is big enough (2.5 to 3 times pump output). Enclose the reservoir if it is an open-top. Remove the old paint and repaint the tank that is compatible to HWCF, or leave it

unpainted. Give a distinct identity to the reservoir meant for HWCF to avoid any misapplication.

2. **Suction lines:** Suction strainer rating should be around 150 μm—no finer—and its capacity should be at least two times the pump flow rate. Remove all bends and narrow passages, if any, to reduce pressure loss. Either an overhead tank or a booster pump must be provided in order to flood the suction side of the pump.

3. **Pumps:** Check the suitability of the pump for HWCF, checking operating speed and pressure with the pump manufacturer.

4. **Valves:** Check with the valve suppliers to see if their products are suitable to HWCF. Change the valve wherever required.

5. **Filters:** Check for the compatibility of filter media with HWCF, as well as the filter rating. Most equipment manufacturers recommend a 10 to 15 μm rating or finer. Add a return line filter if there is not one.

6. **Seals/Metals:** Check for the compatibility of the seals and metals used. Note that this varies from one fluid manufacturer to the other. Follow the manufacturer's recommendation. Replace seals and metals wherever required.

Once the checklist is completed and necessary modifications carried out, the changeover operation can commence.

Changeover

1. Drain all the oil from the reservoir, cylinders, accumulators, and pipelines. Open out the lowest pipeline for effective draining. Clean the tank as thoroughly as possible.

2. Charge the tank with HWCF to the minimum level, taking due care with water quality and mixing.

3. Flush the system with either with no load or minimum pressure. With the flush-fill, operate all functions so that no component in the system is left unflushed. Depending upon the system size, flushing may last two to four hours.

4. While the system is flushing, look for leak spots. Tighten the joints and change the seals as appropriate.

5. Drain the fluid when it is still warm and also without allowing it to settle. Clean the strainer and filter element.

6. Refill the system and operate it at a reduced pressure for some time, then raise the pressure to the designed value.

Post-changeover

1. Closely observe the functioning of the system for a few weeks—encourage both design as well as maintenance personnel to get involved.

2. Arrange for a periodic roundtable meeting of all concerned personnel to exchange their experiences and ideas. Take remedial measures wherever required based on the collective wisdom.

3. Train maintenance personnel to carry out the maintenance schedule and install a procedure to log maintenance data. Periodically review the data collected.
4. Continue this practice till the system is stabilized.

CONCLUSION

With such a host of hassles, why use HWCF? The advantages offered by HWCF are substantial in terms of cost. HWCF is six to eight times cheaper than mineral oil. Since only 5 percent of the fluid must be transported, stored, and disposed of, significant savings can be effected. HWCF has better biodegradability, so environmental and pollution regulations can be easily complied with. Easy availability, excellent fire resistance, low compressibility, and cooler operations are added advantages. HWCF is used successfully in the steel industry, die-casting machines, plate grinders, rolling mills [8, 9, 18], and spot welding robots in the auto industry [20].

As mentioned earlier, if a thorough understanding of HWCF technology and a dedicated maintenance can be ensured, HWCF can contribute sizable savings in direct costs alone. Understandably, these requirements are difficult to meet under shop floor conditions. But the dynamics of technology and the information age may change this situation, and there may be a promising place for HWCF in the future. Then the arduous research and development work that has gone into making HWCF a viable alternative fluid may be adequately compensated.

Chapter 5

FIRE RESISTANT
HYDRAULIC FLUIDS

Fire resistant (FR) hydraulic fluids are used where there is a risk of fire. Molten metals, open flames, electric arc, hot metal surfaces, and high operating temperatures are all potential fire hazards. Welding equipment, die casting and extrusion machines, foundry equipment, forging presses, furnace and glass forming machines, injection molding machines, steel and metal working industries, and mining machinery are some examples of where FR fluids are extensively used. The need for FR fluids was realized in the 1950s, when many fire accidents took place in mining and other industries due to the leakage of high-pressure mineral oil. It should be noted that hydraulic fluid-related fire accidents still take place. Twenty-five people died and 50 were injured in a fire in a food processing factory in the U.S. when a hose burst and pressurized oil sprayed onto a fryer in 1991[1]. To avert such accidents, FR fluids have been made more fire resistant. Also, worldwide concern for safety in the workplace is increasing. Consequently, hydraulically operated devices (instead of people) now perform operations near heat or fire sources, and use of fire resistant fluids is expanding. Tighter government regulations, better understanding of fire hazards and consequent development of test procedures have also made significant contributions toward the development of fire resistant hydraulic fluids. Details on the FR fluids commonly used in industries are covered in this chapter. Any fire resistant fluid must possess the following characteristics:

- The fluid must resist ignition.
- It should have the ability to snuff the flame and prevent it from spreading when the source of ignition is present.
- The fluid must be self-extinguishing when the source of flame is removed.

It is important to realize that FR fluids are neither flameproof nor fireproof; fire resistance means only that these fluids have a higher resistance to ignition than other fluids, and that even if ignited, the fluid has properties that will arrest the spread of flames. The prime reason for using a fire resistant fluid is to reduce the possibility a fire in the workplace. The fluid must also be nontoxic and meet the other technical requirements of a normal hydraulic fluid.

TYPES OF FIRE RESISTANT HYDRAULIC FLUIDS

Broadly, fire resistant fluids can be categorized into two types: aqueous and nonaqueous fluids. Internationally accepted categorizations of FR fluids are given in Appendix 4, Table A4–1. Many fluids are listed in the table, but only the three most widely used fluids are dealt with here (HWCF has been covered in the previous chapter). These fluids are:

- Invert emulsions (water-in-oil-emulsion)
- Water glycols
- Phosphate esters

In terms of the fire resistance of these three fluids and HWCF they are ranked in the following order, beginning with the least fire resistant: water-in-oil emulsions < phosphate esters < water glycols < synthetic solutions < oil-in-water emulsions (HWCF). An overview of FR fluids and their characteristics is given by Phillips [2]. The comparative properties of fire resistant fluids with mineral oil are given in Table 5–1.

Water-Based Fluids

Invert Emulsion (water-in-oil emulsion)   In contrast to oil-in-water emulsion, water is added to oil to form invert emulsion. Invert emulsion belongs to type HFB fluid and is a mixture of 40 percent water and 60 percent oil. Water as tiny droplets in the range of 0.7–2 μm is dispersed into the continuous phase of oil. Water is added slowly into the oil with continuous stirring. In the process, oil surrounds the finely divided water, is dispersed uniformly, and forms a two-phase mixture. Water droplets are kept suspended in the oil by emulsifiers. When ignition takes place, the steam generated blankets the supply of oxygen to the oil and thus prevents the propagation of flame. Invert emulsion is generally available in viscosity grades VG 68 and 100. Data sheets on commercially available invert emulsion are given in Appendix 2, Tables A2–1 to A2–2 [3].

   As the oil content is quite high in invert emulsion, its characteristics are closer to mineral oil; consequently, it is the least fire resistant. The viscosity of invert emulsion is dependent on the viscosity of the oil used, as well as on the percentage of water content. For example, any reduction in water content will reduce viscosity, which can affect the lubrication property; increase in water content will thicken the fluid (contrary to what you might expect). Invert emulsions are non-Newtonian fluids, and this factor should be taken into account when the fluid is used under high shear conditions (for example, in high-speed pumps); viscosity can differ from one part of the system to another depending upon the shear force applied. The ideal operating temperature, i.e., bulk oil temperature, in the reservoir is 50°C (122°F), though 60°C (140°F) is the limit. This

Table 5–1 Comparison of Properties of Fire Resistant Fluids with Mineral Oil

Sl. No.	Property	Mineral oil	Invert emulsion HFB	Water glycol HFC	Phosphate esters HFD
1	Specific gravity @ 15°C	0.88	0.92	1.06	1.15
2	Viscosity @ 40°C cSt	22–100	68–100	22–68	22–100
3	Viscosity index	90–95	140	160	0
4	Pour point °C	−20	−30	−50	−20
5	Flash point (COC) °C	200	–	–	245
6	Water content % volume	–	40	40	–
7	pH	–	–	9	4.2
8	Bulk modulus @ 20°C, bar	2×10^4	–	3.3×10^4	2.12×10^4
9	Specific heat @ 25°C kJ/kg °C	1.89	2.7	3.2	1.6
10	Thermal conductivity @ 25°C W/mK	0.134	–	0.44	0.13
11	Surface tension N/m	23×10^{-3}	–	35×10^{-3}	42×10^{-3}
12	Coefficient of thermal expansion (volumetric)/°C	7×10^{-4}	5.5×10^{-4}	7×10^{-4}	7×10^{-4}
13	Recommended working temperature °C	60	5 to 50	−20 to 50	−20 to 70 Short term 150
14	*Maximum operating pressure, bar (Piston pumps) speed, rpm	240 1,800	150 1,500	170 1,800	22 1,800
15	*Maximum operating pressure, bar (Vane pumps) speed, rpm	200 1,800	70 1,500	150 1,500	200 1,500
16	Autoignition temperature °C	350	–	–	550
17	Cost	X	1.5X	3X	6X

Note: Values given are only typical
*Under best operating conditions

temperature limit takes into account any spot temperature rise of about 15°C to 20°C (27°F to 36°F). A built-in temperature gauge in the reservoir, though not so common in many designs, is a good maintenance aid for controlling the temperature of the fluid. Invert emulsions are nontoxic. Seals compatible with mineral oils are generally also compatible with invert emulsions (Table 5–2).

The main drawback of invert emulsions is their long-term stability in service. The fluids can become either oil-rich or water-rich, thereby reducing their lubricating ability or fire resistant property. Their foaming and deaeration characteristics are not as good as those of mineral oils, hence reservoirs should be adequately sized and provided with baffles to facilitate the collapse of foam and air bubbles. Also, return lines should be located farther from the suction side, and the fluid discharge should be far below the fluid surface so that air entrainment can be prevented.

Table 5–2   Compatibility of Materials and Seals with Mineral Oils and Fire Resistant Hydraulic Fluids

Materials	Mineral oils	Invert emulsion	Water glycol	Phosphate esters
Metals	Compatible with normally used metals	Attack magnesium, lead, and cadmium	Attack lead, zinc, cadmium, magnesium, and non-anodized aluminum; copper alloy may get discolored	Same as mineral oils
Paints	Compatible with most industrial paints	Use only epoxy, phenolic, and nylon-based paints	Same as invert emulsion	Same as invert emulsion
Plastics	Compatible with common plastics	Same as mineral oils	Same as mineral oils	Acrylic, styrene, and PVC are not compatible
Seal materials	Silicone, ethylene, butyl, and propylene are not compatible	Butyl, ethylene, polyurethane, and propylene are not compatible	Except for polyurethane, most materials are compatible	Neoprene, nitrile, Buna N, and polyurethane are not compatible
Other seal materials	Natural rubber not compatible	Leather, cork and natural rubbers are not compatible	Same as invert emulsion	Same as invert emulsion

Note: Data given is based on broad recommendations. Consult also the supplier.

Like HWCF, invert emulsion contains a high percentage of water, so due attention must be paid to engineering the suction lines to avoid cavitations and to promote efficient functioning of the pump. For example, the fluid velocity in the suction side should be below 1m/sec (0.33 ft/sec), the typical value being 0.75 m/sec (0.25 ft/sec), and the maximum vacuum pressure allowed in the suction side is around 7.5 cm of mercury (1.5 psi). The capacity of the strainer should be four times the maximum flow of the pump to minimize pressure loss as well as to effectively control contaminants. The reservoir size should be three to four times the flow rate of the pump. The air gap between the fluid and the reservoir top should be kept as small as possible to minimize loss of water due to evaporation.

The list of applications for invert emulsions cited in published papers is vast [4–9] and includes a host of applications such as the mining industry, foundries, the steel industry and die casting machines. Invert emulsions are inexpensive for many applications, provided the limitations are clearly understood and the fluid is well maintained. Where reduced lubricity and load-bearing capacity are acceptable, and the organization is equipped for the use of water-based fluids, invert emulsion can be cost-effective.

Water Glycols Water glycols belong to type HFC fluids. They are a blend of water, glycol, water-soluble polymeric thickener to give the desired viscosity, and additives to combat corrosion, wear, and foam. The most commonly used glycol is diethylene glycol; other glycols such as ethylene glycol and propylene glycol are also used. The water is mixed with the glycol in a ratio of 40 to 60. Unlike invert emulsion, it is a single-phase solution and a true solution. Water glycol is normally available in viscosity grade VG 46, and sometimes in VG 10 and 22 grades. Because it contains water, its fire resistance is similar to that of invert emulsions. It is a Newtonian fluid, implying that viscosity does not change even under high shear rate conditions. Its viscosity index is very high—around 150—due to the mixture of polymers (not VI improvers). Data sheets on commercially available fluids are given in Appendix A2, Tables A2–3 to A2–6 [3].

The recommended operating temperature is 50°C (122°F). It can tolerate a temperature of 60°C (140°F)—although this is not desirable—if the problem of evaporation is lessened by means of pressurized or closed tanks.

A water glycol's viscosity is dependent on the percentage of water content. Any reduction in water content will increase the viscosity of the fluid, unlike in invert emulsion. An increase in viscosity could cause pump inlet difficulty and a higher-pressure drop in the system, resulting in an increase in fluid temperature. Because it is a water-based fluid, its vapor pressure is low, and hence the inlet velocity of the fluid should be kept below 1 m/sec (0.33 ft/sec), the suction strainer should be adequately sized, and the mesh size should not be less than 350 μm to minimize the pressure drop in the suction lines [10]. The filterability of water glycols is poor compared with that of mineral oils—by at least three times—and accordingly filter size should be three times larger than that one used with mineral oils [10]. Most commonly used seal materials are compatible with water glycols (Table 5–2). They are less harmful to the environment and so disposal is a less serious problem.

Glycol tends to take more deaeration time than mineral oils, hence it should have a higher residence time than mineral oils. The tank therefore should be bigger, with a baffle plate between the suction and return lines. A screen plate in the tank can help air bubbles to burst more rapidly on the fluid surface [10]. Suspended particles in water glycol will take more time to settle, as the specific gravity is higher than that of mineral oil; this calls for a larger reservoir size than that used in a normal mineral oil system.

One advantage of a glycol is its low pour point (–50°C/–58°F), so it poses less of a storage problem and can operate even in severe winter conditions without any problem. Due to their low pour point, water glycols are a favored choice, where the ambient temperature is low, for example, in mining applications in Europe. Because of its poor lubricating property, its load-carrying capacity with respect to rolling bearings is not satisfactory, and in such applications use of glycols should be carefully examined.

Water glycols are commonly used in steel industries, aluminum and zinc die casting industries, forging and extrusion machines, glass forming machines, welding machines, and mines [11–16].

Nonaqueous Fire Resistant Fluids

To overcome the inadequacies of mineral oils and water-based fluids, many synthetic fluids have appeared on the market since the late 1920s, though not all of them are meant for fire resistant property requirements. Phosphate esters, polyol esters, silicones, silicate esters, polyphenyl ethers, polyalphaolefin, and fluorocarbons are the most widely used synthetic fire resistant fluids. Except for phosphate esters, all other fire resistant synthetic fluids come under the ISO type HFDU. Among them, except for phosphate esters and polyol esters, the usage of all other fluids is largely confined to aerospace and military applications rather than in any industrial applications. In terms of fire resistance, these synthetic fluids are ranked in the following order, beginning with the least fire resistant: polyalphaolefin < polyol esters < silicate esters < silicones < polyphenyl ethers < phosphate esters < fluorocarbons.

Polyol esters are still used in many industrial applications, for example, in industrial robots and welding equipment. Low specific gravity; high VI; and good thermal, hydrolytic, and oxidation stability are its advantages. In recent times polyol esters have gained better acceptance because they are biodegradable. In other words, they are fire resistant as well as biodegradable. Another key advantage is that conversion from mineral oils to polyol esters is very easy, as they are compatible with mineral oils and with many metals and seals that are used with mineral oils. But their fire resistance is only marginally better than that of mineral oils, and polyol esters do not pass all fire resistance tests. Hence their usage is now limited to applications where the fire resistance requirement is less stringent. They are two to three times costlier than mineral oil. The most widely used nonaqueous fire resistant fluids are phosphate esters, and these are discussed in detail in the succeeding paragraphs.

Phosphate Esters Phosphate esters are nonpetroleum and nonaqueous products that are chemically synthesized; they come under the ISO classification of an HFD-R fluid. The most commonly used phosphate esters, particularly in industrial applications, are based on triaryl phosphates. They are made by the reaction of phenol with phosphorous oxychloride, as given below [18]:

Ar.OH	+	POCl$_3$	→	(ArO$_3$)PO	+	HCL
Phenolic material		Phosphorus oxychloride		Triaryl phosphate		Hydrochloric acid

Sometimes they are also referred to as synthetic phosphate esters, mainly to differentiate them from tar acid phosphates or natural phosphates. Due to the reduced availability of coal tar, natural phosphate esters are rare, and phosphate esters are now mainly synthetically produced. Consistency in composition and quality of synthetic types are easily maintained. However, they are less hydrolytically stable than natural phosphate esters, and hence they more easily become acidic. Data sheets on commercially available phosphate esters are given in Appendix 2, Tables A2–7 to A2–9 [3].

Phosphate esters are inherently fire resistant because of the fire extinguishing property of the phosphorus molecule present in the fluid [18]. The fuel value of these fluids is 25 to 30% less than that of petroleum fluids, and as such their ability to sustain their own combustion is weak [19, 20]. Their operating temperature is higher than that of mineral oils; it can go up to 150°C (300°F) for short periods of time. Anticorrosion, antifoam, and the lubricating property of phosphate esters are comparable with that of mineral oils. Phosphate esters are polar and can easily form chemically adsorbed film on the surface, which makes a metal surface impervious to water. Hence they are inherently rust resistant and seldom require fortification with anticorrosion inhibitors. As these fluids have good lubricating property over other FR fluids, phosphate esters, can be used for the lubrication of bearings in hydraulic systems. But their VI value is low unless fortified with VI improvers.

The specific gravity of phosphate esters is the highest among hydraulic fluids, and they demand more power to pump the fluid than the other fluids do. Heat transported by a fluid is dependent on the product of specific gravity and specific heat, and hence the higher density of phosphate esters to some extent offsets the lower value of specific heat. However, under the same operating conditions, phosphate esters will show a marginally higher fluid temperature than mineral oil.

Commonly used seal materials such as neoprene or nitrile, and the paint used with mineral oils, are not compatible with phosphate esters (Table 5–2). They are also not compatible with mineral oils, and in fact mineral oil is considered a contaminant. Furthermore, some manufacturers generally do not recommend that their fluid be mixed with phosphate esters of another make because composition can differ. Phosphate esters are free of many drawbacks associated with water-based fluids, and in this respect they can be treated like petroleum products. Modern phosphate esters are free of toxicity and are biodegradable [21,22].

A discouraging factor is their direct cost; phosphate esters are about six to eight times more expensive than mineral oil. Because of their high cost, phosphate esters are generally selected only where demanding lubrication and load-bearing requirements must be met. Phosphate esters are quite suitable in applications that use water-based fluids—to name a few, hot metal ingots, casting machines, furnace controls, metal working industries, rolling mills, robots, and welding machines. Their applications in turbines, however, seem to be universal [23–25]. Though their initial cost is high, in smaller systems, this may not be a significant deterring factor when compared with the advantages of phosphate esters. Also, in some applications, they may be found to be economical in the long run because of their long life and less demand for maintenance.

SELECTION OF FIRE RESISTANT FLUIDS

There are no hard-and-fast rules for the selection of fire resistant fluids. Each fluid has its own merits and demerits. In terms of fire resistance, lubrication, cost-effectiveness, operating temperature, maintenance, load capacity, operating pressure, and eco-acceptability, each fluid offers a balance of varying performance requirements. For example, in terms of the best possible protection against fire hazards, cost, and ecological acceptability, water-based fluids score over the others. But these benefits come at the cost of poor lubricating property, lower operating temperature, and unfavorable maintenance requirements. So with water-based fluids, there is always a trade-off between fire resistance and lubrication performance. Nonaqueous fluids like phosphate esters, on the contrary, are better suited where good lubrication, higher operating temperature coupled with good protection against fire risks, and lower maintenance demands are necessary. Apart from technical considerations such as fire resistance, lubrication requirements, and compatibility with hydraulic equipment, many nontechnical factors such as prevailing maintenance practice, management policy, size of the system and consequent cost, and in some cases conversion cost may also influence the selection of the fluid. However, one criterion that should override all other factors—including cost—is that the selected fluid should provide the best possible protection against all possible fire hazards in given operating conditions.

Past practices also influence the choice of fluid. Invert emulsions, for example, are commonly used in the U.S. and the U.K., but they are not favored in Germany [16, 26], particularly in coal mining. Water glycol is the widely used fluid in European mines. In other words, there is no universality in the choice of fluids. Borowski explains [5] how he made invert emulsion work successfully in a steel slab caster application when no one was using it in such an application. The systematic way he tackled all the technical problems and the pains he took to make the application work prove that with an open mind, new solutions can be explored. On the other hand, for a similar steel slab caster application, Fletcher favored water glycol, based on his experience [15]. Do these examples remind us of the old adage "One man's meat is another man's poison"? There is always more than one solution. All factors must be considered; this sometimes requires exploratory tests/test runs and follow-ups, which many organizations are less inclined to encourage. Also not many people like taking such pains. Yet each system must be closely examined and carefully assessed in terms of operational requirements, fire safety requirements, desired component life, and modifications required to the system. No pain, no gain.

CHANGEOVER PROCEDURES

Changeover procedures must be well planned and executed, as the properties of FR fluids are drastically different from those of mineral oils. Guidelines for changeover are available from many sources [27–40]; national standards and manufacturers' technical bulletins are good sources.

Mineral Oil to Invert Emulsion/Water Glycols

Preliminary Groundwork First, make the operating personnel familiar with the properties of water-based fluids as compared with mineral oils, which will eliminate many operating problems. Check the suitability of all the hydraulic elements, such as pumps and valves, for the given operating conditions, e.g., speed, pressure, and particularly for the high density and low viscosity of these fluids. As water glycols are poor lubricants, the hydraulic components and rolling bearings used in the system must be carefully examined for the suitability of these fluids. Check particularly the compatibility of all the seals, filter media, metals, and paints (Table 5–2). Seek also the recommendations of the fluid supplier. Make changes wherever appropriate. The checklist given for HWCF is also applicable to these fluids.

Draining and Flushing Drain the system completely—tank, cylinders, accumulators, and all pipelines. Disconnect and open out the lowest piping point to ensure effective draining. Thoroughly clean the tank as well as possible. Blow all the lines with dry compressed air. Fill the tank to the minimum level with the fluid. Flush the system without any load for about 30 minutes. Because mineral oil is not miscible with glycol, it will be floating at the top and can be skimmed off. The allowable residual oil for HFC fluid should be below 0.1 percent. Any increase in mineral oil content will result in the formation of deposits as well as considerable reduction in the life of the fluid [14]. Watch for any leakage; tighten the joints or change the seals/gaskets as appropriate. Inspect the strainer and filter; if necessary, change them. Drain the system when the fluid is still warm and without allowing it to settle. Discard the flush fluid if it is found to be very contaminated; otherwise it can be used for flushing operations.

Refilling and Commissioning Refill the system with fresh fluid and put the system into operation. Slowly increase the load to the designed value. Monitor the system operation for two weeks; particularly inspect the strainer and filter. Because water glycols have detergent action, initially they may load the strainer and filter. If necessary, change them; strainer/filter clogging can cause cavitation in the pump. Once the system is stabilized, install the maintenance procedure and monitor the parameters that are given in Chapter 8. Record the data, review it periodically, and take corrective measures as appropriate at the right time.

Mineral Oil to Phosphate Esters

Preliminary Groundwork Train all operating personnel about the properties of phosphate esters, particularly their high cost and the care they require. Check the suitability of all the components in the system for its compatibility and suitability with the phosphate esters (Table 5–2). Seals that are used with mineral oils are generally not compatible with esters. Take into account the higher specific gravity of the fluid while examining the suitability of the pump and filters. Seek also the recommendations of the fluid supplier with regards to compatibility. Make changes wherever appropriate.

Draining and Flushing Drain all the fluid in the system thoroughly as explained earlier, and clean the tank as well as possible. Fill the tank to the minimum operating level with phosphate esters prefiltered to 3 μm level. Install a 100 mesh strainer and auxiliary filter to increase the filter capacity and improve the removal of contaminants. For effective flushing, the system may be divided if possible. Phosphate esters have a solvent effect and hence may carry more solid particles than mineral oil, necessitating more flushing time. FMC Corporation, a leading supplier of phosphate ester fluids, recommends that flushing be carried out in turbulent flow condition so that the internal surfaces of the pipe and equipment can be effectively cleaned [35]. The recommended flow rate should be greater than 0.143dv liters/minute where d is the bore diameter in mm and v is the fluid viscosity in cSt at flushing temperature between 55°C and 60°C (130°F and 140°F). Take care to protect the hydraulic elements, such as orifice valves, gears, or cylinders, which cannot withstand turbulent conditions. Monitor the mineral oil content and particulate level. The mineral oil content should be 0.1 percent or less. When the level of desired cleanliness is reached, flush-fill can be drained off while it is still warm. Blow out the residual fluid with clean, *dry air*. Dry air is emphasized here because moisture/water is highly undesirable to phosphate esters.

Refilling and Commissioning Charge the system with fresh fluid after pre-filtering it with a 3 μm filter. Change the strainer and filters. Run the system initially with reduced pressure and slowly increase the pressure to the designed value. Monitor the levels of mineral oil, water content, and particulates once a week until a stabilized condition is observed. Monitor the parameters that are outlined in Chapter 8 and record the data. Take corrective measures as appropriate at the right time.

Chapter 6

TEST METHODS ON FIRE RESISTANT HYDRAULIC FLUIDS

Over the years, test procedures for mineral hydraulic oils have been standardized and agreed to by user and manufacturing industries as well as by the various standardization bodies. But common test standards to evaluate the fire resistance of hydraulic fluids are still a distant goal. The fact that there are approximately 50 test methods [1] and more than a dozen standardization bodies involved indicates the complexity of the problem. In fact, agreement even on the terminology related to "fire resistance" is still in the evolving stage. The reasons are not far to seek. A wide variety of ignition sources—for example, the many combinations of fluid leakage and ignition in industrial practice—have brought about many types of ignition tests. Also, it is difficult to simulate the exact fire conditions encountered in different applications. Fire hazard conditions in mining industries differ considerably from those in, say, metalworking or aerospace industries; molten metal ignition tests meant for hot metal casting operations may not adequately represent the operating conditions of metalworking industries. In other words, each industry has its own particular fire hazards that require their own set of test conditions, and hence it is difficult to develop a universal test procedure. These tests, therefore, at best can be used to compare the relative flammability of different fluids; they cannot fully ensure the performance of a fluid in actual fire conditions, which can differ very much from those of the test.

Given the complex nature of the problem, the efforts to develop more reliable and acceptable methods are still continuing. Extensive work on test methods has been done at various centers [1–13], and we can be optimistic that, in the near future, the number of test procedures will narrow, signifying that users, manufacturers, and standardization bodies have reached a consensus on fire resistance.

It is beyond the scope of this book to give an account on all the test procedures that are available; only those that are important and commonly used in industries are outlined here. Most of the tests are qualitative, i.e., the results they give are pass/fail. Quantitative assessments, of course, would help users discriminate among available fluids and select the one that is most suitable for the given operating conditions. Summaries of common test procedures are given in references [1–4].

Broadly, the tests for fire resistant fluids can be classified into two types: fire resistant property tests, and non–fire resistant property tests.

Fire resistant property tests	Non–fire resistant property tests
• Autoignition temperature test	• Anticorrosion test
• Hot surface ignition test	• Vapor-phase rust test
• Wick ignition test	• Emulsion stability test
• Spray ignition test	

FIRE RESISTANT PROPERTY TESTS

It may be recalled from the previous chapter that any fire resistant fluid should resist ignition and prevent propagation of flames. Most tests are designed to assess one of these properties. Flash point and fire point, which are also considered fire resistant properties, have been discussed in Chapter 3. Other test procedures that can be termed "core fire resistant tests" are detailed in the following paragraphs.

Autoignition Temperature Test

This test is widely used to determine the temperature at which ignition occurs spontaneously under standard test conditions. It is an important fire resistant property and is commonly quoted as a parameter in specifications of fire resistant fluids. But a test procedure that is acceptable to everyone concerned is still elusive. Two ASTM standards on autoignition temperature (AIT)—D 286 and D 2155—were withdrawn; the new ASTM standard E 659[14] is not specific to hydraulic fluids but applies to any liquid chemical. Formulating an acceptable test standard for this property is still under close review. The two test procedures that are commonly used are given here: the DIN and the U.S. Bureau of Mines methods.

DIN 51794 Method This test procedure is based on German standard DIN 51794:1978 [15]. The apparatus used is shown in fig. 6–1. The test is conducted in air at atmospheric pressure, at a predetermined temperature in the absence of a flame under standardized conditions. First, the ignition temperature is estimated as a predetermined coarse approximation so that the rate of heating the ignition chamber can be controlled to a finer degree as the temperature approaches the autoignition temperature. Sample fluid is injected by hypodermic syringe drop by drop, as slowly as possible, into a temperature-controlled, heated 200 mL open Erlenmeyer conical glass flask containing air. Autoignition is indicated by the distinguishable appearance of a flame or detonation inside the flask within a maximum ignition delay of five minutes. Ignition delay is the time between the introduction of the sample and the appearance of flame. At least three of the lowest values from the main determination and the repetition

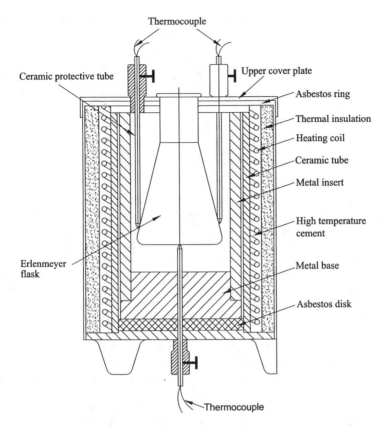

FIGURE 6-1 Apparatus for Autoignition Temperature (DIN Method)

must fall within a span of 10°C (18°F) for temperatures up to 300°C (572°F), and within a span of 20°C (36°F) for temperatures above 300°C. The lowest internal flask temperature at which autoignition occurs for a series of prescribed sample values (not mean values) is the autoignition temperature of the fluid in air at atmospheric pressure. The final value is rounded off to multiples of five (rounded off to the lower). The procedures for the preparation of test apparatus and samples, the safety precautions to be observed, and how to conduct the tests in series of trials is quite elaborate; for details the reader may refer to the standard.

U.S. Bureau of Mines Method This test method, formulated by the U.S. Bureau of Mines [16], has also been adopted by the Indian Standard [24]. It is specifically meant for the coal mining industry. In principle, the test apparatus and the procedure are similar to those of DIN. However, there are great differences in the construction of the apparatus and in the details of the test procedure; readers may refer to the standard for an in-depth look.

The U.S. Bureau of Mines method injects the sample differently, and also arrives at time lag and ignition temperature differently. Unlike the DIN method,

Table 6–1 Comparison between AIT and Flash/Fire Point for Known Hydraulic Fluids and Solvents [1]

Fluid/solvent	Autoignition temperature °C ASTM D 2155	Flash/fire point (open cup) °C ASTM D 92
Mineral oil ISO VG 46	350	200/225
Phosphate ester ISO VG 46	575	245/360
Toluene	536	4/ *
Acetone	519	-19/ *

*Fire point data is seldom quoted for solvents. Usually they are close to the flash point and determined by closed cup procedure.

0.07 mL of test sample is injected in full into the heated 200 mL borosilicate glass Erlenmeyer test flask with a hypodermic syringe, and the syringe is withdrawn immediately. The time that elapses between the instant of injection and that of ignition, as evidenced by the appearance of flame, is noted as time lag. If there is no flame in five minutes after injection of the test sample, the sample is considered nonflammable at the test temperature. If ignition occurs within five minutes or less, the test is repeated by reducing the temperature of the test flask in decrements of 5°F until ignition no longer occurs, and this temperature is noted as the first ignition test temperature for the 0.07 mL test sample. The test is repeated with a sample of 0.10 mL. If the ignition temperature observed is lower than that obtained with the 0.07 mL sample, the test is repeated with samples of 0.12 mL, 0.15 mL, and so on, in increments of 0.03 mL, until the lowest ignition temperature is observed.

In contrast, if the lowest temperature observed with a 0.10 mL sample is greater than that obtained with 0.07 mL, then the test is repeated by reducing the sample to 0.05 mL and then to 0.03 mL, until the lowest ignition temperature is observed. The fluid is not considered fire resistant if the observed ignition temperature of the sample under the test described above is less than 600°F (315°C). Unlike the DIN procedure, the observed ignition temperature is not quoted as an absolute value but is used to classify whether the fluid is fire resistant or not.

Autoignition temperatures are not quoted for water-based fluids. It should be noted that the measured value is highly dependent on the sample size and on the volume of the glass flask—for example, a bigger flask will give a lower ignition temperature. Hence, when a comparison is made, the type of equipment used and the test conditions should be taken into account.

In their detailed study, Kuchta and Cato [17] point out that there are many variables that can cause variations in the test results: sample mass, vapor pressure of the sample, injection pressure, fluid atomization, shape and vessel size, and vessel material. When so many variables can influence and distort the test results, skepticism about how accurately this test assesses fire resistance is justified. Philips reports [1] that it is a widely misunderstood and misquoted test procedure. His data (Table 6–1) indicates that according to the AIT assessment, highly volatile substances like toluene and acetone not only qualify as fire resistant fluids, they are more fire resistant than mineral oils! He cautions that AIT values should not be viewed in isolation when assessing a fluid's fire resistance; other methods of assessment must also be considered.

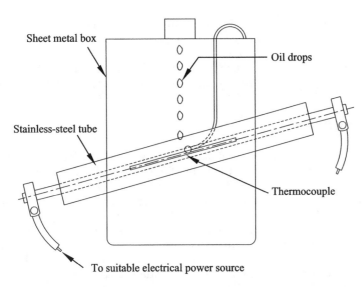

Sheet metal box

Oil drops

Stainless-steel tube

Thermocouple

To suitable electrical power source

FIGURE 6–2 Schematic Diagram of Hot Manifold Test (AMS 3150 C) (Reprinted from SAE AMS 3150 C © 1986 Society of Automative Engineers, Inc.)

Hot Surface Ignition Test

Leaked oil falling on a hot surface can cause ignition; this is one of the most common causes of fire. The hot surface ignition test procedures commonly used are the hot channel test and the hot manifold test. These tests are used for both aqueous and nonaqueous fluids.

Hot Channel Test (Factory Mutual Research) In the hot channel test [18], formulated by Factory Mutual Research, U.S. per the Standard 6930, a length of steel channel inclined at 30 deg to the horizontal is heated by a gas burner to 704°C (1,300°F). The heat source is removed, and the fluid is sprayed onto the steel at a pressure of 6.9 MPa (1000 psi) and a temperature of 60°C (140°F) from an oil burner nozzle positioned at 150 mm (6″) from its surface for 60 seconds. If the fluid burns for less than five seconds after the spray is removed, the fluid passes the test.

Hot Manifold Test (AMS 3150C) The SAE's AMS 3150C [19] hot manifold ignition test (fig. 6–2) represents the leakage of fluid onto a hot surface, and it also indicates the tendency of a fluid to propagate flame.

Ten mL of oil is slowly poured (instead of sprayed) onto a stainless steel tube 76 mm in diameter and 1 mm thick (3″ O.D. × 0.04″ wall), heated electrically to 704°C (1,300°F) and kept at an angle of ≈ 5 deg to horizontal from a specified height in not less than 40 seconds. The test is repeated several times from different heights and onto different parts of the tube. The temperature of the tube is kept constant by thermostatic control, unlike in the channel test where the surface temperature is not controlled. As the fluid is poured, any flash or burning on contact with the tube, or whether the flame is carried to the fluid collected at the bottom of the shield, is noted. The flammability is recorded thus: a) the fluid burns on the tube b) it does not burn on the tube c) it burns, flashes, or does not burn in the bottom of the shield.

This test procedure as an ISO document is under way: ISO 20823:2001, *Petroleum and related products—Determination of the flammability characteristics of fluids in contact with hot surfaces – manifold ignition test* [20].

Wick Ignition Test

The wick ignition test measures flame propagation and persistence of burning. It is a pass/fail test and is not suitable for aqueous fluids. The test is useful in determining the flammability of different fluids; it is used to evaluate the flammability of a fluid dispersed in an absorbent material and is not suitable for evaluating highly flammable materials. This test is of interest mostly to mining industries, particularly the coal-mining industry. It can also be applied to power-generating equipment, where fire can result from oil leaking into insulation or lagging material used in steam pipes. A variety of test methods are available, but only the commonly used procedures are described here.

ISO 14935: Method One method is the recently introduced ISO test method [21]. It is used to assess the persistence of a flame applied to the edge of a wick of nonflammable material immersed in a fire resistant fluid (fig. 6–3a and fig. 6–3b). Because the test relates to the bulk behavior of a fluid, the results may provide useful information for its safe transportation and storage. This test procedure does not apply to aqueous fluids.

A nonflammable aluminosilicate board 195 mm long, 25 mm wide and 2 mm thick (7.67″ × 1″ × 0.08″) is soaked in the test fluid and then placed in a reservoir 200 mm long, 25 mm wide and 20 mm deep containing the test fluid; one edge of the board is left exposed (fig. 6–3a). A propane flame from a nozzle 0.6 mm in

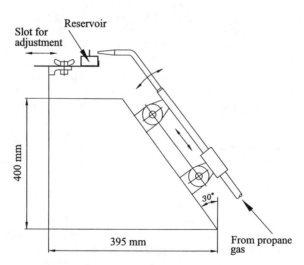

FIGURE 6–3a   Schematic Diagram of Wick Test Apparatus (ISO Method)

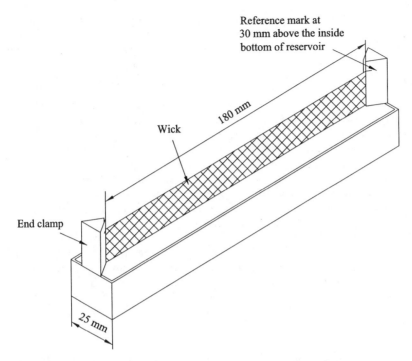

FIGURE 6–3b   Wick Test–Reservoir (ISO Method)

diameter is applied to the midpoint of the exposed edge of the aluminosilicate board for two seconds. The persistence of the flame after the removal of the igniting propane flame in seconds (nearest to 0.1 seconds) is noted. The test is repeated six times for each of the five different periods of flame application, namely, 2 seconds, 5 seconds, 10 seconds, 20 seconds, and 30 seconds. The persistence of these five different periods of application of the flame to the board is calculated per the procedure given in the standard, and the largest of these averages is taken as the result. It is really a pass/fail test; however, the fluid is considered fire resistant if the absolute maximum value does not exceed 60 seconds. The test is conducted at an ambient temperature of 18°C to 22°C in still air. Details on the test apparatus are given in the standard.

ASTM D 5306 Method Per the ASTM D 5306 [22] method, the linear flame propagation rate on a ceramic fiber impregnated with the test fluid under specific conditions is determined. This test is used to compare the relative flammability of fire resistant hydraulic fluids. A section of the ceramic fiber is impregnated with the sample fluid under specified conditions and is then placed on a standard support with a 50 gram weight attached to both ends of the fiber to provide necessary tension. A differential thermocouple pair is kept 2 mm above the fiber and is connected to a strip chart recorder. The fiber is ignited near one end of the support. The flame is allowed to advance along the length of the fiber past each thermocouple until it extinguishes itself upon reaching the opposite fiber support. The differential thermocouple and the recorder determine the time required for the flame front to propagate across a measured distance. The average propagation rate is then calculated from the measured distance of flame travel, as is the time required for the flame front to propagate over that distance. Many test results indicate that phosphate esters exhibit no flame propagation under this test.

Flame Propagation Test in a Mixture of the Fluid and Coal Dust (7th Luxembourg Report, Section 3.2.2) As stated in the introduction of this chapter, each type of industry has its own particular fire hazard requirements, and its test method is formulated to meet those requirements. This test in this section is exclusively used by coal mining industries, and it is also like a wick test, except that the fluid is soaked in coal dust instead of in a wick. The test is not simulated with other materials. The commonly followed test procedure is the one formulated by the European Community Mines Safety Commission [23]. Per this method, 75 grams of coal dust is mixed with 37.5 mL of test fluid (2:1 ratio); from this mixture, test pieces of 150 mm long and 20 mm thick are prepared. The test piece is placed on a steel plate and heated by a gas burner for five minutes. The tip of the burner is kept at least 45 mm from the underside of the steel plate. The following points are noted:

- The greatest distance traveled by the tip of the flame, expressed in mm
- The time it takes for flame to die out
- Any anomalies, such as glowing after extinction of flame, or extinction followed by renewed ignition

The results are expressed as an arithmetic mean of the 10 tests. A flame travel of not more than 10 cm from the point of ignition in an average of 10 tests is taken to be the general limit [10]. For details on the test apparatus and test piece preparations, readers may refer to the standard.

This test is more suitable for nonaqueous fluids, as the water content in the soaked coal could evaporate under actual operating conditions. Also, it should be noted that although specified coal dust is used in the test, any variation in this parameter could become a source of dispersion in the test results.

Pipe Cleaner Test (U.S. Bureau of Mines) Another test that falls in this category is the pipe cleaner test followed by the U.S. Bureau of Mines [25]. In this test, an ordinary pipe cleaner 50 mm (2″) long is soaked in the test fluid under specific conditions for two minutes and attached to an oscillating mechanism such as a windshield wiper. The pipe cleaner is oscillated in the horizontal plane 25 cycles per minute and made to enter and leave the flame of a standard Bunsen burner. This cycle is repeated until a self-sustaining flame is observed on the pipe cleaner. The number of cycles required to obtain a self-sustaining flame is noted. The test is conducted with the test fluid under three conditions: 1) the test fluid at room temperature 2) the test fluid heated for two hours to a temperature of 150°F (65.5°C) and then cooled to room temperature 3) the test fluid heated for four hours to a temperature of 65.5°C and then cooled to room temperature. Under each condition, five tests are conducted. For the fluid to qualify as fire resistant, the following conditions must be met:

1. The average number of cycles before attaining a self-sustaining flame under the test fluid condition one shall be 24 or more.
2. The average number of cycles before attaining a self-sustaining flame under the test fluid condition two shall be 18 or more.
3. The average number of cycles before attaining a self-sustaining flame under the test fluid condition three shall be 12 or more.

It may be noted that the test does not relate to any possible fire hazard. Also, the test is viscosity-dependent to the extent ISO VG 100 mineral oil can pass this test, and the test conditions are more favorable to aqueous fluids [1].

Spray Ignition Test

This is another category of test to evaluate the fire resistant property of a fluid; it is important because the test conditions are closer to those of a possible fire hazard—a leaking pipe, a hydraulic line rupture, or a punctured hose under high pressure, all of which can produce a spray. Pressurized fluid spraying on hot metal during operating conditions is a certain fire hazard. The spray ignition test assesses both the fluid's ignitability as well as its flame propagating tendency. The U.S. Bureau of Mines test, the 7th Luxembourg Report tests, and the Factory Mutual Research spray ignition test are the commonly quoted test methods, which will be dealt with in the following paragraphs.

U.S. Bureau of Mines Test The U.S. Bureau of Mines formulated this test [26] for use in coal mining industries, and it has also been also adopted by the Indian Standards [24]. The test assesses the flammability characteristics of hydraulic fluid by spraying the fluid onto three different sources of ignition.

A 2.5-quart (~2.5 liter) sample fluid is heated in a pressure vessel to a temperature of $150 \pm 5°F$ (65.56°C), and nitrogen is introduced into the pressure vessel at 150 psi (10.2 bar). The heated fluid along with the pressurized nitrogen is sprayed through an atomizing round spray nozzle with an orifice 0.025″ (0.635 mm) in diameter at an angle of 90 deg at the following three ignition devices:

1. A metal trough with a metal cover in which cotton waste is soaked in kerosene
2. An electric arcing device in which the arc is produced by a 12,000-volt transformer
3. A propane torch-Bernzomatic or equivalent

The fluid is sprayed at each ignition device, which is moved along the trajectory of the spray. The ignition device is kept at different distances from the nozzle tip for one minute or until the flame or arc is extinguished (if less than one minute). The fluid is considered fire resistant if:

• The test does not result in ignition of any sample fluid, or the ignition of a sample does not result in flame propagation for a time interval not exceeding six seconds at a distance of 18 inches (457 mm) or more from the nozzle tip to the center of each igniting device.

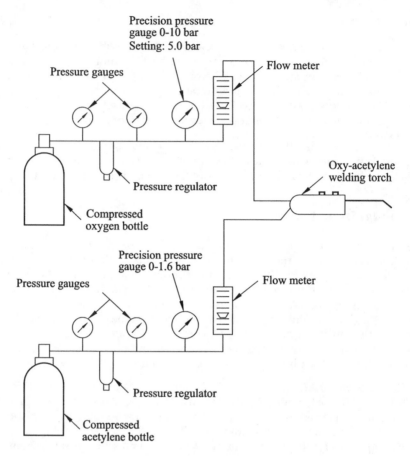

FIGURE 6–4 Schematic Diagram of Ignition Torch (U.K. Spray Ignition Test)

7th Luxembourg Report Tests The 7th Luxembourg Report contains three test procedures. The first is the "Community of Six Spray Ignition Test" per section 3.1.1, which is used mainly in continental Europe. The second, often called the "U.K. Spray Ignition Test", is described in section 3.1.2 [27]. The third method, described in section 3.1.3, is known as the "Stabilized Flame Heat Release Spray Test". When sufficient experience has been gained with this new method, the first two conventional methods will be discontinued. Therefore, only the second and the third methods are described here.

U.K. Spray Ignition Test In the U.K. test (fig. 6–4), the test fluid is sprayed onto only one ignition device. The test fluid is assessed for its fire resistnce by spraying the fluid under a pressure of 70 bar (1,015 psi) on to a precisely defined oxygen/acetylene flame. After ignition of the spray, the igniting flame is withdrawn, and the persistence of the spray's burning is recorded. The fluid is heated to a test temperature of 65°C (149°F) for aqueous fluids and 85°C

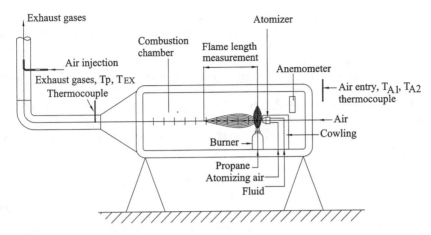

FIGURE 6–5 Schematic Diagram of Spray Test (7th Luxembourg Report Test)

(185°F) for other fluids. The heated fluid is sprayed from an 80° hollow–cone spray oil-burning nozzle, at a pressure of 70 bar (by pressurizing it with nitrogen from a cylinder) onto an oxyacetylene flame from a welding blowpipe. When the burning of the spray is established, the igniting flame is repeatedly applied and withdrawn from different positions along the length of a pressurized spray of fluid, and the time of the persistence of burning after each spray, i.e., the time between removal of the igniting flame and extinction of the combustion of the spray, is noted. The highest time recorded is the spray flame persistence. The ignition criterion of the spray jet and the flame length is related to the post-combustion of the spray after the removal of the igniting flame. For the fluid to pass, the maximum persistence of burning in the spray after withdrawal of the igniting flame should not exceed 30 seconds. Details on the test apparatus (fig. 6–4) are outlined in the report. The recently released ISO standard [28] is based on this test procedure.

Stabilized Flame Heat Release Spray Test The test procedure given in section 3.1.3 of the 7th Luxembourg Report (fig. 6–5) is more severe and elaborate than the U.K. spray ignition test. This test uses a steady input of heat to ignite the spray and to produce a stabilized spray flame. The burning spray is said to be stabilized when its energy release, length, and other properties are steady as a function of time. The heat release is determined by the temperature of the exhaust gases, and this is related to the ignitability of the fluid. This test involves elaborate calibration of test apparatus and test preparations, and readers may refer the standard for details.

The test fluid at a rate of 90 mL/minute and at a temperature of 13°C to 20°C (55°F to 68°F) is sprayed horizontally, by means of a stainless steel pneumatic atomizer, along with compressed air into a combustion chamber 2000 × 490 × 490 mm (79″ × 19″ × 19″) and exposed to a defined propane flame from a gas burner. The compressed air is supplied at the rate of at least 1.92 Nm³/hour (m³/hour at normal temperature and pressure. i.e., NTP) and at a pressure of around 2 bar (29 psi) and a temperature of 10°C to 25°C (50°F to 77°F). The propane gas premixed with air produces a continuous vertical flame for igniting the spray. The temperatures of the test chamber, fresh air, and exhaust gases are accurately measured when the burner is ignited, both without and with fluid coming out of the nozzle, with compressed air coming out of the nozzle in both cases. The temperature recorded is used to calculate the ignitability factor RI on the basis of the relationship between the difference in temperatures of the exhaust gases and fresh air when just the burner is ignited, as well as when the test fluid mixed with air is burning with a stabilized flame. RI is given by the equation:

$$RI_W = [500\,(T_p - T_{A1})]\,/\,[7(T_{EX} - T_{A2})]$$

Where RI_W – ignitability factor; T_p – exit temperature °C; T_{A1} – air entry temperature without jet spray °C; T_{EX} – temperature of combustion gases with jet spray °C; T_{A2} – air entry temperature with jet spray °C.

The calculation procedures are detailed in the report. All measurements are made under well-defined, stable conditions, as prescribed in the test procedure. Initially the test is conducted with a propane flow rate of 0.13 Nm^3/hour, and if the calculated value of RI is greater than 50, then the test is repeated with a propane flow rate of 0.4 Nm^3/hour. In this case, RI $_W$ is calculated from the following equation:

$$RI_W = [100\,(T_p - T_{A1})\,/\,(T_{EX} - T_{A2})] + 30$$

However, there is only one grading scale without specifying the propane flow rate. On the basis of RI values, the fluids are ranked for ignitability. The fluids are graded from A to H, where A refers to the safest fluid, e.g., HWCF and H, to the most hazardous ones, e.g., mineral oils. The ignitability RI and the resulting ranking are the main factors used to classify fluids. Along with the measurements made to determine the value of RI, flame length and smoke opacity are also measured On the basis of flame length and smoke opacity, secondary rankings of the fluids are possible. These secondary rankings are not used to classify a fluid as either pass or fail; they are used as a guide for a specific application. Fluids that pass the U.K. test generally have an RI in excess of 25.

Factory Mutual Research Spray Ignition Test Another notable test procedure [29–32] in this category is the one formulated by Factory Mutual Research, which is commonly used in the U.S. In this test (fig. 6–6), the hydraulic fluid at a temperature of 60°C (140°F) is sprayed vertically upward through a conical nozzle with an exit diameter of 0.38 mm. The nozzle is located 1.5 m (59″) from the floor, in the center of a propane-air burner with a diameter of 0.14 m (5.5″) and a capacity of 15 kW (20 HP), with the flame height at about 0.2 m (7.87″). The tip of the nozzle is kept in the same plane as the ring burner head. The test is conducted at a pressure of 6.9 MPa (1,000 psi). The combustion products along with the ambient air are collected in the sampling duct, where various parameters such as flow rate of fire products and air, gas temperature, generation rates

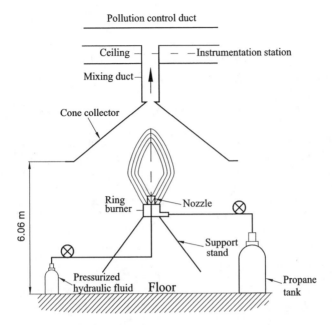

FIGURE 6–6 Factory Mutual Research Spray Test Setup

of carbon monoxide (CO) and carbon dioxide (CO_2), and the consumption rate of oxygen (O_2) are measured to calculate the chemical heat release rate from the generation rate of CO and O_2.

Spray flammability by itself may not be useful, and this method addresses the problem by ratioing ignition energy (critical heat flux) to energy release measured in the test. A nondimensional parameter known as the spray flammability parameter (SFP), which is related to the critical heat flux required for fuel ignition and the chemical heat release rate of the burning process, is used as an indicator of the burning rate, i.e. the flame propagation rate. SFP is defined as:

$$SFP = 4\,Q_{ch} / (\pi d_s \times q''_{cr})$$

where Q_{ch} is the chemical heat release rate in kW; q''_{cr} is the critical heat flux for ignition below which ignition is not possible, expressed in kW/m^2; and d_s is the equivalent diameter of the nozzle expressed in meters (nozzle exit diameter multiplied by the square root of the fluid density at the test temperature to the ambient air densities in kg/m^3). q''_{cr} is related to the fire point temperature, $T_{fire}(K)$ of the fluid, as follows:

$$q''_{cr} = \alpha\, \sigma\, T^4_{fire}$$

Where α is the fluid surface absorptivity (assumed as unity), and σ is the Stefan-Boltzmann constant ($5.67 \times 10^{-11}\ kW/m^2K^4$). A detailed account of the derivation of the equation is given in references 27–29. Based on the value of SFP, hydraulic fluids are classified as follows:

Group 1: SFP ≤ 20×10^4; the fluid will be unable to stabilize a spray flame
Group 2: SFP > 20×10^{4} but < 40×10^4; the fluid is less flammable than mineral oil
Group3: SFP > 40×10^4; flammability is closer to mineral oil.

More recently, Factory Mutual Research has "normalized" the calculation of SFP by dividing by the mass flow rate of the fluid during the spray flammability test, and has appropriately adjusted the ranges for these groupings [32]. This compensates for different flow rates achieved in different test apparatuses, as the flow rate is not controlled, only measured. The new range boundaries now are defined at 4 and 8×10^4, rather than at 20 and 40×10^4. From this it follows that fluids with a higher SFP value have a higher flame propagation rate, and that their flame will stabilize more easily than those of fluids with lower SFP values.

The spray ignition test is applicable to all types of fire resistant fluids, and thus the fire resistance of all fluids can be measured and compared. Despite the fact that every test method has its own limitations, the spray test is closer to reality than other test methods.

SELECTION OF TEST METHOD

The wide varieties of test procedures outlined thus far should give readers concise insight into the nuances of possible fire hazards, the difficulties encountered in replicating them, and also the tests' limitations and applicability to actual applications. No single test can be said to represent completely any particular application. Every application must be closely examined, and the possible fire hazards must be perceived from many angles. In a hot metal molding operation (tilting crucibles, die-casting, furnace door operation, etc.), for instance, a hose bursting near molten metal could be a potential fire hazard. The fluid to be used under such conditions probably should satisfy the hot manifold test as well as one of the spray ignition tests. Some industries may have norms; for example, the fluids that are used in mining and other exacting applications must satisfy one of the spray tests and one of the flame propagation tests, per the European Union norm. [12]. A clear understanding of the test procedures, measurements, the interpretation of test results, and the tests' limitations in simulating actual fire conditions is of paramount importance before selecting the type of test for a specific application. Broad guidelines for selecting the relevant test method from the fire resistant test procedures outlined above are given in Table 6.2 [1].

NON–FIRE RESISTANT PROPERTY TESTS

Non–fire resistant property tests such as pour point, antirust, copper corrosion, foaming tendency, antiwear, seal compatibility, hydrolytic stability, and water

Table 6–2   Suggested Test Procedures for Specific Fire Test Properties [1]

Property	Test method	Test standard
Ignitability	Hot manifold/hot channel test	AMS 3150 C/Factory Mutual Research 6930 tests
Self-propagation	Wick ignition test; hot channel test or hot manifold test	ISO 14935/ASTM D 5306; AMS 3150 C/Factory Mutual Research 6930 tests
Ignition delay	Hot manifold/hot channel test	AMS 3150 C/Factory Mutual Research 6930 tests
Smoke	Spray ignition test	7th Luxembourg Report 3.1.2 and 3.1.3/Factory Mutual Research spray ignition tests
Heat release	Spray ignition test	7th Luxembourg Report 3.1.2 and 3.1.3/Factory Mutual Research spray ignition tests
Relevance to major hazards	Spray ignition, wick ignition, hot manifold/hot channel tests	7th Luxembourg Report 3.1.2 and 3.1.3/Factory Mutual Research spray ignition tests; ISO 14935/ASTM D 5306/AMS 3150 C/Factory Mutual Research 6930 tests
Ability to evaluate all fluid types on an equal basis	Spray ignition, hot manifold/hot channel tests	7th Luxembourg Report 3.1.2 and 3.1.3/Factory Mutual Research spray ignition tests; AMS 3150 C/Factory Mutual Research 6930 tests

content are also related to fire resistant fluids and have been covered in the Chapter 3. Therefore only the tests specific to fire resistant fluids, i.e., anti-corrosion, vapor-phase rust prevention, and emulsion stability are described here.

Anticorrosion Test (ISO 4404)

This test is used to assess the corrosion protection characteristic of water-based fire resistant fluids, namely HFA, HFB, and HFC types on five metals conforming to one of the given standards, i.e., steel (EN grade C45/AISI 1045), copper (ISO grade Cu ETP or Cu-FRHC/ASTM grade C11000), brass (ISO grade Cu Zn 35/ASTM grade C 26 800), zinc (ISO grade 99.5), and aluminium (ISO grade Al 99.5/1050A)—the materials normally used in hydraulic systems and installations. Based on the European standard Comité Europeen des Transmissions Olehydrauliques et Pneumatiques (CETOP), a new ISO standard [33] has been formulated, which is detailed here.

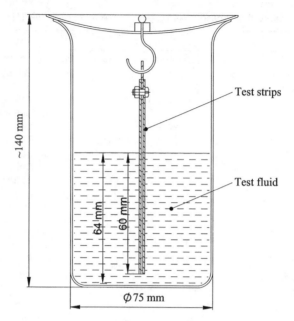

FIGURE 6–7 Anticorrosion Test Setup with a Pair of Test Strips (ISO Method)

The test setup is given in fig. 6–7. A test specimen measuring $100 \times 20 \times 1$ mm ($4'' \times 0.8'' \times 0.04''$) is suspended inside a glass beaker of 400 mL capacity and immersed to a depth of about 60 mm in the test fluid (250 mL). A single test strip is kept in each of the five beakers, and four pairs of strips—steel/zinc, copper/zinc, zinc/aluminium, and aluminium/steel—are kept in another four beakers; a tenth beaker is kept without any test strip for comparison purposes, i.e., to evaluate the change in the fluid. All 10 glass beakers are kept inside a thermostatically controlled heat bath. The temperature of the test fluid is maintained at 35°C (95°F). All the beakers are covered with a watch glass to minimize fluid evaporation losses. The test specimens are kept in this state undisturbed for a period of 28 days (672 hours) without any interruption. After 28 days the test strips and the test fluid are inspected and assessed for the following:

Test strips:

- General appearance
- Deposits formed

Test fluid:

- Color
- Appearance
- Deposits

Table 6–3   Rating System for Metal Strips and Test Fluid

Rating	Metal strips	Test fluid
0	No effect	No effect
1	Slight color change or oxidation of less than 20% of the surface	Deposits
2	Strong color change	Separation, e.g., surface accumulation of oil, or distinct phase separation
3	Deposits or oxidation on more than 20% of the surface	Cloudiness of an initially clear fluid
4	Corrosion or pitting	Color change
5	Other effects to be specified	Other effects to be specified

The metal strips and test fluid are rated as given in Table 6–3. After thorough cleaning and drying of the metal strips, each one is weighed to the nearest one mg. The rating and the difference in mass as plus or minus are reported, depending on the increase or decrease in the mass. In addition, the following factors may also be noted:

- No corrosion appears on the metals.
- The mass loss should not exceed 10 mg per test strip for materials with a density of more than 7 kg/dm^3 and 5 mg for materials with a density of less than 7 kg/dm^3; the mass loss for Zn should not be more than 20 mg, and for aluminum not more than 10 mg.
- No deposit should form on the test strip, and mass increase should not be more than 5 mg per test strip.
- No significant color changes should be observed on the metal strip.

Vapor-Phase Rust Prevention Test (ASTM D 5534)

As outlined in the HWCF, water-based fluids are always fortified with corrosion inhibitors to protect against corrosion in both the liquid and vapor phases. This test method is used to assess the ability of a hydraulic fluid to prevent rusting of steel due to vapors of hydraulic fluid and water. The ASTM test procedure [34] is the one generally followed, and it has two parts—Part A is used for water glycols and HWCF (water in continuous phase); Part B is used for invert emulsions and mineral oils.

A steel test specimen 30.2 mm (13/16″) in diameter and 4.8 mm (3/16″) thick conforming to Grade 1018 of specification A108 or BS: 970-1955 EN 3B is attached to the underside of the cover of a beaker containing the test fluid (fig. 6–8). Next, 275 mL of test fluid is poured into the beaker and placed in the oil bath, which is maintained at a temperature of 60°C (140°F). The oil level in the bath is maintained such that it is not below the oil level in the test beaker. The stirrer is set in the test beaker so that the temperature of the test fluid reaches 60°C. For Part A of the test, the test specimen is kept for six hours from the time the temperature reaches 60°C.

For Part B of the test, the test fluid is stirred for 30 minutes after its temperature reaches 60°C (140°F). Then the beaker cover is raised, and the water

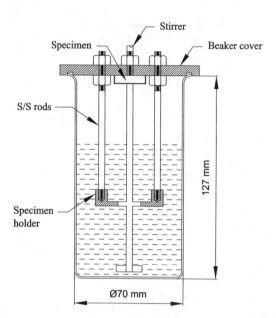

FIGURE 6–8 **Schematic Diagram of Vapor Phase Rust Apparatus (ASTM Method)**

beaker containing 25 mL of deionized water is placed in the undercarriage and the beaker cover assembly is replaced as for Part A of the test. Under this condition the specimen surface is exposed to water in the vapor state. The test specimen is kept in this state for six hours with the stirrer in continuous operation. After six hours, the test specimen is removed and inspected. If the test specimen shows any rust on the horizontal test surface that faced the test fluid, under normal light without any magnifying glass, then the fluid fails. The test is conducted in duplicate. The fluid is reported as a pass only if both specimens meet the pass criterion.

Emulsion Stability Test

As the invert emulsion is a two-phase fluid, it has a tendency to separate into its primary phase, i.e., water and oil, which can happen during storage as well as during operation. This separation phenomenon is known as the fluid's instability, and it can happen either due to the poor quality of the fluid or to improper mixing. Obviously this test is applicable only to type HFB fluid, namely, water-in-oil emulsion (invert emulsion). There are three test procedures—two per ASTM D 3709 and D 3707, and the third per IP 290. The IP method is more commonly used in the U.K. and continental Europe.

ASTM D 3709 Method Per the ASTM D 3709 procedure [35], the stability of the emulsion is tested when it is subjected to temperature cyclic changes between ambient and −18°C (0°F). The significance of this test is to assess the ability of the emulsion to withstand severe winter conditions under storage. A

100 mL sample contained in a 100 mL measuring cylinder is kept in a cooling box at −18°C (0°F) for 16 hours and then allowed to stand at a room temperature of 21°C (70°F) for eight hours. This procedure is repeated for three more cycles. After the fourth cycle, the sample is kept in a cold box for 64 hours and then for eight hours at ambient temperature. The cycle is repeated for four more cycles, and in the final cycle the sample is kept at room temperature for only three hours. After the end of test cycle, the sample is examined and the following are reported:

1. The amount of water separated in % volume
2. The amount of oil separated in % volume
3. Upper layer water content in % weight
4. Lower layer water content in % weight.

ASTM D 3707 Method The ASTM D 3707 procedure [36] focuses on the emulsion's stability during usage. The stability of the emulsion is tested at 85°C (185°F). A 100 mL sample is placed in a temperature-controlled oven at 85°C for 48 hours for Procedure A, and 96 hours for Procedure B. At the end of 48 or 96 hours, the sample is kept at a room temperature of 21°C (70°F) for one hour; the sample is then examined and the same four parameters indicated in ASTM D 3709 are reported. After 48 hours, per the Denison HF-3 specification [37] there should not be any free water, free oil should not exceed 4%, and water content in the upper and lower portions must not vary more than 5%.

IP 290 Method The IP method [38] measures the separation of water and oil, i.e., the phase separation and settlement of dispersed water that occurs during storage at ambient temperature, and gives an arbitrary measure of the extent to which the dispersed water undergoes settlement. A sample of 600 mL emulsion is taken, and volume percentage of water is determined per the procedure prescribed in IP 290/ASTM D 95 or any equivalent standard. Next, 450 mL of emulsion is poured into a 500 mL measuring cylinder over a layer of 50 mL of distilled or deionized water. The cylinder is sealed and allowed to stand undisturbed for 1,000 hours at room temperature. After 1,000 hours, the volume of surface oil layer and the change in volume of water layer, i.e., the volume of accumulated free water, are read. The water content of these samples is determined and compared with the measurement of initial water content. The following parameters are reported along with the maximum and minimum ambient temperature and standing time:

1. The surface oil layer volume
2. The accumulated free water volume
3. The change in water content at the upper 425 mL level
4. The change in water content at the lower 125 mL level

A positive value indicates an increase in water content, while a negative value indicates a decrease in water content. Most of the emulsions exhibit no free water, and free oil is generally within 5 ml. The change in water content at the 425 mL and 125 mL level is normally around 1%.

Depending on the operating conditions of a given application, users can select the applicable test from the three procedures just described.

Chapter 7

BIODEGRADABLE HYDRAULIC FLUIDS

Biodegradable fluids are a recent addition to the family of hydraulic fluids. Growing industrialization and its impact on environmental degradation due in part to the misuse and mishandling of lubricants has prompted European nations, Germany in particular, to formulate performance criteria for biodegradable hydraulic fluids and also to evolve standards for biodegradability [1, 2]. In Germany alone, per 1992 estimates, 50 million liters (13.2 million gallons) of mineral oils per annum were dumped into the drain water and ground, and 27 million liters (7.13 million gallons) of hydraulic oil disposed of into the surroundings [3, 4]. Today, those figures are probably much higher. The situation in the U.S. and other countries is similar. Per the National Fluid Power Association (NFPA) data, of 830 million gallons of industrial lubricants used in the U.S., only 75 million gallons are recycled—hardly 9% [5]. The rest, it seems, are dumped into the environment. The impact of mineral oil polluting the environment is detailed in Chapter 9. Work is under way to address this problem before it becomes unmanageable; the basic approach seems to be "prevention is better than cure." So as a first step, efforts are being made to reduce—if not eliminate—the usage of mineral oils. In this direction, steps have been taken to end the use of mineral oils in equipment that operates in water or near water sources and other environmentally sensitive areas. In addition, a range of biodegradable hydraulic fluids are now commercially available.

Biodegradable lubricants made their first appearance in 1980 to meet the needs of two-stroke engine oils. Ester-based fluids were first used in hydraulic systems in the late 1980s [6]; but vegetable-based fluids have also been in use since the 1980s. Applications of biodegradable lubricants have since expanded and include engine, gear, turbine, and slide way lubricants; metal cutting fluids; steel rolling oils; engine coolants; textile loom oil; and mould release oils [7, 8, 31]. Yet despite their wide range of availability, these fluids are still not widely used. In Germany, for example, annual consumption of these fluids [4] is estimated to be only 2.7 million to 4.5 million liters, which is indeed a small market share. A survey conducted by the British Fluid Power Association [9] also suggests that usage of these fluids is currently limited to hydraulic systems installed in highly sensitive watercourses where leaks or accidental spillage of mineral oil can contaminate soil, water, or groundwater. Biodegradable fluids have yet to make a significant mark in the industrial applications, and this is apparently due to the absence of either legal or technical compulsions. Admittedly, biodegradable fluids do not perform as well as other lubricants in use today. Nevertheless, given the importance of environmental protection and the demands of the future, users of hydraulic fluids must become aware of and understand the technicalities of biodegradable hydraulic fluids. Hence, a concise overview on this topic is outlined in this chapter.

BIODEGRADATION: WHAT DOES IT MEAN?

There is yet no universally accepted definition of biodegradation, as it is a complex mechanism still not fully understood. The generally accepted explanation of biodegradation, however, is this: Biodegradation is a biochemical process by which a substance decomposes both in soil and water by natural processes brought about by the metabolic action of living organisms or their enzymes. This means that biodegradable substances can be digested or consumed by microorganisms present in the water, air, or soil, and that during that process no vegetation or aquatic organisms are killed, or harmful or toxic substances left behind.

Biodegradation is an essential and potent natural process of the life cycle. In fact, decay of any substance is a natural process. The diverse microbial population present in the environment and the ability of these microbes to use a wide range of organic compounds help decaying compounds, even complex ones, decompose into simple compounds, which are ultimately assimilated as biomass—biomass is nothing but a living organism or part of a living organism. It is this biodegradation process that recycles carbon, oxygen, nitrogen, phosphorous, etc., into the biosphere. If it weren't for this biodegradation process, the earth would perish of its own waste. Biodegradation is obviously essential to a clean environment.

CLASSIFICATION OF BIODEGRADATION

Broadly, there are two classes of biodegradation [3, 10, 11]. The first is primary biodegradation, which involves the conversion of the original substance into new chemicals through biological action; the second is ultimate biodegradation, which involves the complete conversion of the original substances into carbon dioxide, water, new microbial biomass, and simple indigestible inorganic substance. This process is also sometimes referred to as mineralization. From this it follows that biodegradability of a substance can be measured by means of assessing the following parameters—loss of organic carbon, oxygen consumed, formation of carbon dioxide and water, or a combination of these. The biodegradation process is illustrated in fig. 7.1 [12].

There are also two other classifications known as readily biodegradable and inherently biodegradable. A substance is classified as one of these depending upon the percentage degradation of the substance measured per the appropriate

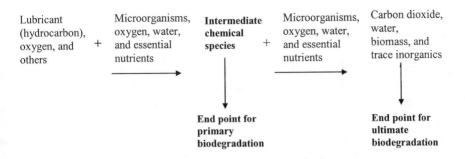

Note: 1. Typical essential nutrients include nitrogen, phosphorous, sulfur, calcium, etc.
2. Intermediate chemical species may be hydrocarbon with shortened chain length, acetic acid, glyceride, amines, phenols, etc., depending upon the composition of the fluid.

FIGURE 7–1 Biodegradation Process

test procedure (Table 7–1). Per the U.S. Environmental Protection Agency (EPA) definition, ready biodegradability describes those substances which in certain biodegradation test procedures produce results that are unequivocal, and which lead to the reasonable assumption that the substance will undergo rapid and ultimate biodegradation in aerobic aquatic environments [42].

Inherently biodegradable means that certain substances can degrade under favorable circumstances, such as aerated conditions, oxygen-rich conditions, etc. In other words, the test aims to assess whether the substance has inherent potential for biodegradation. For example, contrary to the general thinking, most synthetic fluids and mineral oils are not nonbiodegradable but are classified as inherently biodegradable, i.e., they can biodegrade under favorable aerobic conditions. In a CEC (Coordinating European Council) test, the percentage degradation for mineral oil was found to be as high as 45 percent, though it could be as low as 10 percent [3]. This difference is due to the variation in the structure of hydrocarbons. For example, increasing chain length and branching tend to decrease biodegradability. Many tests reveal that oxygen is a prime nutrient in biodegradation.

The parameter measured and the pass level against the various test procedures are given in the Table 7–1. These tests measure either oxygen consumption or carbon dioxide production under specified environmental conditions, normally over a period of 28 days. Biodegradation is done by exposing the test substance to an inoculum—spores, bacteria, single-celled organisms, or other live materials [13]. It's important to note that these tests differ from one another in that the source of inoculum, its concentration and size, aeration method, and parameters measured are different in each test. Hence they are not technically equivalent. Depending upon the application, tests must be chosen.

It is quite surprising that highly biodegradable material like vegetable matter buried in a landfill does not degrade even after 20 years, possibly because of anaerobic conditions and lack of water [10]. This suggests that biodegradation processes have certain prerequisites. Biodegradation of hydraulic fluids is a

Table 7–1 Biodegradability Criteria for Various Test Methods

Test method	Days	Test environmental conditions	Suitable for	Parameter measured	Pass level*	Test for
OECD 301A-1992 (ISO: 7827-1984) DOC Die-away test	28	Aquatic, aerobic, dark, or diffused light @ 22°C	Water-soluble and nonvolatile substances	Loss of dissolved organic carbon	>70%	Ready biodegradability
OECD 301 B-1992 CO_2 evolution test (modified Sturm test)	28	Aquatic, aerobic, dark, or diffused light @ 22°C	Water-soluble, poorly soluble, and nonvolatile substances	Production of CO_2	>60%	Ready biodegradability
OECD 301 C-1992 Modified MITI test (1)	28	Aquatic, aerobic, dark @ 25°C	Water-soluble, poorly soluble, and volatile substances	Oxygen consumption	>60%	Ready biodegradability
OECD 301 D-1992 Closed bottle test	28	Aquatic, aerobic, dark @ 22°C.	Water-soluble, poorly soluble, and volatile substances	Oxygen demand	>60%	Ready biodegradability
OECD 301E-1992 Modified OECD screening test	28	Aquatic, aerobic, dark or diffused light @ 22°C	Water-soluble and nonvolatile substances	Loss of dissolved organic carbon	>70%	Ready biodegradability
OECD 301 F-1992 Manometric respirometry test	28	Aquatic, aerobic @ 22°C	Water-soluble, poorly soluble, and volatile substances	Oxygen consumption	>60%	Ready biodegradability

Test	Days*	Conditions	Substances	Measurement	%	Biodegradability
OECD 302 A-1981 Modified semi-continuous activated sludge test (SCAS)	>28	Aquatic, aerobic,	Water-soluble and nonvolatile substances	Loss of dissolved organic carbon	>20%	Inherent biodegradability
OECD 302 B-1992 Zahn-Wellens/ EMPA test	28	Aquatic, aerobic, dark, or diffused light @ 22–25°C	Water-soluble and nonvolatile substances	Loss of dissolved organic carbon	>20%	Inherent biodegradability
ASTM D 5864-1995	28	Aquatic, aerobic, dark @ 20–25°C	Water insoluble and nonvolatile substances	Production of CO_2	–	Ready biodegradability
US EPA shake flask test	28	Aquatic, aerobic, dark @ 20–25°C	Water-insoluble and nonvolatile substances	Production of CO_2	–	Ready biodegradability
CEC-L-33-A-93	21	Aquatic, aerobic, dark @ 25°C	Poorly soluble and nonvolatile substances	Loss of hydro-carbon	>67%	Primary biodegradability
				Infrared bands	>80%	

*Percentage biodegration level
CEC–Coordinating European Council; OECD– Organization for Economic Co-operation and Development
MITI–Ministry of International Trade and Industry, Japan
EMPA–Swiss Federal Laboratories for Material Testing and Research

function of many factors. The chemistry of the fluid and the physical, chemical, and biological conditions of the environment under which it will undergo biodegradation—factors like temperature, quantity of oxygen, bacterial population (numbers and variety), viscosity, and water solubility—greatly influence the speed at which the fluid can biodegrade.

TYPES OF BIODEGRADABLE FLUIDS

Triglycerides, synthetic esters, and polyglycols are the three base fluids currently used to formulate biodegradable hydraulic fluids. As in other types of hydraulic fluids, various additives enhance and fulfill the other technical requirements of a normal hydraulic fluid. Details on these fluids are given in the succeeding paragraphs.

Triglycerides

Triglycerides are water-insoluble fluids whose main constituents are natural fats and oils from either vegetable or animal sources. They are also called natural esters and come under the ISO classification of HETG fluids.

The chemical composition of triglycerides consists mainly of a variety of saturated fatty acids such as lauric, myristic, palmitic and stearic acids; monounsaturated fatty acids such as oleic and erucic acids; and polyunsaturated fatty acids such as linoleic and linolenic acids, which are esterified to a glycerol backbone. Rapeseed oil, for example, contains as many as 12 types of fatty acids, all of which have markedly different melting points [14]. The physical properties depend on the composition and concentration of these fatty acids present in the fluid. A higher level of saturated fatty acids may affect the fluid flow characteristics at low temperature; a higher level of polyunsaturated fatty acids may affect the oxidation resistance property. Variation in weather conditions can also cause variations in the constituents of seeds, which makes it difficult to get consistent desired performance characteristics from triglycerides.

This problem can be dealt with in different ways [15–26]. One is to compensate for triglycerides' inadequacies by fortifying them with suitable additives. Another possibility is to devise a suitable refining technology that retains only the desired type of fatty acids, monounsaturated types, for example. Or go to the genesis of the seed: Modify its genes so that the oil contains only certain fatty acids, and then blend the oil with additives. Extensive work on additives to biodegradable fluids is already under way, particularly for rapeseed and esters, the notable contributions coming from the Lubrizol Corporation in the U.S., and Rhein Chemie Rheinau GmbH and Institute for Fluid Power Transmissions and Control (IFAS) in Germany [16–21]. The usual additives are corrosion inhibitors, antioxidants, and pour point depressants. Additives must be compatible with the environment (i.e., biodegradable) as well as with other co-additives; this is a technically complex task [16–21]. Additives that are used for mineral oils are generally not suitable for biodegradable fluids, as they do not meet the toxicity and biodegradability criteria. The refining technology for vegetable oils is still in an evolving stage. [15, 22].

Genetic modification may sound like loud thinking, but biotechnology is an emerging technology. DuPont has developed a genetically modified soybean with high oleic content; genetically modified soybean oil gives better oxidation stability and antiwear performance [23–26, 51].

The fluids presently marketed under the triglycerides category are mostly made from rapeseed oil. Soybean-, sunflower-, and canola-based oils are also used but in a limited way, particularly in the U.S. and Canada [27–30]. Because rapeseed oil is plentiful in Europe, it is the most commonly used base fluid for triglycerides there. The annual crop of soybean oil is about 24 billion pounds in the U.S. [24]. This abundance of soybean oil plus its cheaper price may make it an alternative to rapeseed oil, at least in the U.S. Work to improve the performance of soybean oil is continuing. The fluid stability problem associated with rapeseed has been addressed, and commercial formulations already exist. In the case of soybean oil, problems of inadequate performance have yet to be resolved to satisfy commercial requirements, although genetic modification of the seed is a promising start.

Performance Characteristics Triglycerides are available in viscosity grades VG 22, 32, 46, and 68; the viscosity-temperature characteristic of these fluids is very good, as their viscosity index is generally more than 200. Because their molecular weight is twice that of mineral oils, the vapor pressure and volatility of triglycerides are low, and hence their flash point is higher than that of mineral oils, which means they are more fire resistant than mineral oils. Since triglycerides are also made up of carbon and hydrogen, physical properties such as specific gravity, specific heat, and bulk modulus are close to that of mineral oils. As these fluids already contain oxygen in their structure, they are easily biodegradable; triglycerides are in fact rated the most biodegradable. However, because their biodegradability can deteriorate during operation, care must be taken to prevent leakage.

The high fat content of these fluids makes them good lubricants; it also gives them poor oxidation and thermal stability characteristics. Consequently their operating temperature is lower and operating life shorter than mineral oils. Their antiwear performance, can match that of mineral oils, and the FZG gear test damage class is found to be more than 10—even with no additives, they can pass the load stage of 10, compared with the load stage of six achieved by mineral oils with no additives [14]. Extensive vane pump tests conducted by Denison Hydraulics, France, also confirm that triglycerides and synthetic esters demonstrate better antiwear performance than antiwear mineral oils [31]. They also readily pass the Vickers vane pump 35VQ25 test (Chapter 3). Test results indicate that triglycerides possess good antifoaming characteristics as well.

Triglycerides are sensitive to water, as their hydrolytic stability is poor; hence water content in these fluids should be limited to less than 0.1 percent. But poor hydrolytic stability means easy biodegradability—a dilemma—because water ingressed into these fluids aids to break the ester linkage, i.e., splits the fluid into alcohol and acids, and thus initiates the biodegradation process. Triglycerides are miscible with mineral oil, but a high content of mineral oil reduces their ecological acceptability.

Table 7–2 Recommended Seal Materials for Biodegradable Fluids (ISO VG 32, 46, and 68) per VDMA Standard 24569

Fluid Temperature	<60°C	<80°C	<100°C	<120°C
HETG	AU	AU*	–	–
	NBR	NBR	–	–
	HNBR	HNBR	–	–
	FPM	FPM	–	–
HEES	AU	AU*	–	–
	NBR*	NBR*	–	–
	HNBR*	HNBR*	–	–
	FPM	FPM	FPM	FPM*
HEPG	AU*	–	–	–
	NBR*	NBR*	–	–
	HNBR	HNBR	HNBR	HNBR
	FPM	FPM	FPM*	FPM*

VDMA- German Association of Machinary and Engineering
Note: Recomendations given are applicable for VG 32, 46, and 68 grades only.
*seals that are subjected to predominantly dynamic loads needs investigation for each application
HNBR–Hygrogeneted NBR
FPM–Fluorinated rubber
AU–Polyurethane
NBR–Nitrile-butadiene rubber
(Courtesy: The British Fluid Power Association, Extract from *Report BFPA/P 65-1995*)

The flow characteristic of triglycerides under low temperature is peculiar compared with mineral oils. As stated earlier, the fluidity at lower temperature depends upon the type of fatty acids and their concentration in the fluid [1, 14, 20]. Apart from the chemical structure of triglycerides, physical conditions such as their age and the temperature at which they are stored also significantly change the low temperature flow characteristic. In other words, this property is a function of temperature and time. A fluid that has a pour point of −24°C under warm conditions may drop to −12°C if it is stored at 5°C for a week [1]. Bush and Blacke have observed that fractional crystallization and precipitation also influence this behavior [20]. In undoped rapeseed oil stored at −10°C, micro crystallization was observed after 12 hours and large precipitations after 36 hours; the oil was fully crystallized and solidified after 60 hours. But when the oil was heated to room temperature, it reverted to its original state. When the oil was fortified with additive, it became cloudy after 60 hours at the temperature of −20°C. If a system must operate under low temperature conditions for a long period, consultation with the manufacturer is recommended. Factors such as how long the fluid will be stored and the ambient temperature of the storage place must also be considered. The upper limit for the operating temperature is quoted as 80°C; however, at this temperature, triglycerides are prone to rapid oxidation, viscosity increase, and obnoxious odor. Therefore it is advisable to limit the operating temperature to 70°C [32–33].

Triglycerides have good compatibility with nitrile and viton seals; their compatibility with seals is also dependent on the operating temperatures and the extent of the fluid degradation (Table 7–2). High operating temperatures and highly degraded fluid can make the compatibility of these fluids with elastomers uncertain. Users should carefully examine the compatibility of a fluid with the various materials used in the system for the given operating conditions (Table 7–4).

Triglycerides are currently used mostly in mobile hydraulics, forestry machinery, construction and agricultural machinery, hydraulic lifts and elevators, and harvesting machines, when operating temperatures and loads are moderate. They are also preferred where the volume of fluid and the chance of leakage are high [35]. It is quite encouraging to note that triglycerides are also used in machine tools as hydraulic fluids, gear oils, and cutting fluids [31]. Such varied applications indicate a promising future for biodegradable fluids.

Synthetic Esters

These are also water-insoluble fluids and are a reaction product of organic acids and alcohols. Unlike triglycerides, which are natural products, synthetic esters are chemically synthesized. Hence there are many propriety products available with a combination of many chemical types. Base oil properties such as viscosity, oxidation stability, and pour point can be varied by selecting different alcohols and acids, which can give a wide range of base oils that satisfy a variety of applications. Consequently, performance characteristics can differ significantly from one product to another. The chemical structure of synthetic esters is similar to that of natural triglycerides. Synthetic esters come under the ISO classification of HEES fluids. The most commonly used synthetic esters are made from tri methylo propantrioleate (TMP-oleate), tri methylo propane-esters (TMP-esters), and carboxylic esters such as di isotridecyl adipate (DITA). Combinations of high oleic vegetable base oils blended with biodegradable synthetic esters are also commercially available (best of both worlds), and they meet many of the coveted OEM specifications such as Denison HFO; Cincinnati P-68 & 70, and Vickers I-286-S & M-2950-S [7, 34].

Performance Characteristics Synthetic esters are highly biodegradable, as they already contain oxygen. They are easily attacked and broken down by microorganisms—they decompose into alcohols and acids, which are easily digested. However, biodegradability can vary from one fluid to another, depending upon the chemical composition. For example, the biodegradability of unsaturated synthetic esters (TMP-oleate) is better than that of saturated ones (TMP-esters).

Synthetic esters are available in viscosity grades VG 22, 32, 46 and 68, and their viscosity index varies from 150 to 190. The base fluid characteristics are very good and hence the additives required to enhance their properties are minimal [4]. It is an added advantage, as it is only the additives, which interfere with the biodegradability of the fluid. Their thermal and oxidation stability are better than those of triglycerides, and their antiwear performance can match that of mineral oils. Synthetic esters are more expensive than triglycerides, but they give better service life than triglycerides do because of their better oxidation stability; to a large extent, this offsets their higher cost. The new generation of saturated esters possess the same oxidation stability as that of mineral-based oils [6, 7]. In fact, synthetic esters are gaining ground as some of the best biodegradable hydraulic fluids. As the application of synthetic esters expands, their price may become competitive with that of other hydraulic fluids.

Synthetic esters are miscible with mineral oils, but mixing them with mineral oils, can affect their filterability, foaming tendency, and compatibility with metals [36]. These fluids also suffer from poor hydrolytic stability, though not to the same degree as triglycerides, and their water content should be limited to less than 0.1 percent. As in triglycerides, hydrolytic action can split esters into acid and alcohol and initiate the process of biodegradation. The autocatalytic hydrolytic dehydration phenomenon observed in phosphate esters is also applicable to synthetic esters. Hence keeping these fluids at a low TAN level could keep hydrolysis in check. As this is a major performance weakness of synthetic esters, efforts have been made to overcome this problem by blending the fluids with a suitable additive. By adding sterically hindered carbodiimide, this problem has been overcome [21]. Extensive laboratory tests indicate that the additive reacts with the acids and transforms them into stable and neutral urea derivatives, which arrest the acceleration of ester splitting. The urea-type reaction products are soluble in esters. Another advantage of this additive is that by itself it does not get hydrolyzed for a long period of time and is also eco compatible.

The pour point and operating temperatures of synthetic esters are higher than those of triglycerides—they can tolerate a temperature range of $-20°C$ to $90°C$. A distinct advantage of synthetic esters over triglycerides is their good performance under low temperature conditions. Synthetic esters with a pour point of $-50°C$ to $-60°C$ are commercially available [37]. Their compatibility with seals and various materials is generally similar to that of triglycerides (Tables 7–2 and 7–4). But possible variations in operating conditions and in the chemistry of one ester to another can significantly affect the compatibility of seals and other materials [36]. It is advisable to check this aspect with the fluid and equipment manufacturers for a given application.

Synthetic esters are a preferred choice over triglycerides where the application demands higher load bearing and operating temperatures. Typical applications include forest machinery, snow crawlers, wheel loaders, and excavators [35]. In fact, it is claimed that synthetic esters are suitable even for machine tool applications as well as for the lubrication of moderately loaded gears [31, 37].

Polyglycols

Unlike triglycerides or esters, polyglycols are water-soluble fluids and many types of glycols can be formulated. But the currently used ones are polyethylene and polyalkylene glycols. They come under the ISO classification of HEPG fluids, are available in viscosity grades VG 22, 32, 46, and 68, and their viscosity index is lower than that of triglycerides or synthetic esters—around 150 only.

The usage of polyglycols as lubricants is not something new: they are found in food industries, textile and rubber lubrication, compressor lubrication, etc.

Performance Characteristics The compressibility of polyglycols is lower, but their density is higher than that of mineral oils. Hence pump inlet conditions and response time of relief valves need careful engineering. Good shear stability and aging characteristics, plus comparable antiwear performance with mineral oils, are other positive points of polyglycols. Their oxidation stability is comparable to that of mineral oils, and hence oil-change intervals are almost the same as for mineral oils. An added advantage is their fire-resistant property. Foaming tendency, however, is not as good as that of mineral oils. The incompatibility of polyglycols with various materials is one distinct negative aspect compared with tryglycerides and synthetic esters: Common coatings, seals, and gasket materials are not compatible with polyglycols. Viton seals or hydrogenated nitrile seals are safer to use. Seal compatibility is also dependent on viscosity, i.e., a seal compatible with VG 46 fluid may not be suitable for VG 32 grade [38].

Polyglycols are not miscible with mineral oils, and the mineral oil content should not exceed 1 percent of the total volume. Their operating temperature ($>80°C$) is higher than that of triglycerides and synthetic esters. Because of their good miscibility with water, polyglycols are generally used in equipment that operates near water sources, such as lift bridges and locks, dredges, swimming pools, and washing installations [38]. They have also been used in printing, lime kiln, and beverage industries in Germany [39]. They are much more expensive than tryglycerides.

Polyglycols are ethers and contain oxygen in their structures, which aids biodegradation. Their biodegradability, however, is inferior to that of the other two fluids; their biodegradability rating depends upon the type of glycol they contain—for example, polyethylene glycols are more easily biodegradable than polypropylene glycols. The molecular weight of polyglycols also influences the rate of biodegradability, and hence a high-viscosity fluid has a slow rate of biodegradation. Also, because of their easy miscibility with water, polyglycols can easily seep into groundwater or mix with river water; if sufficient oxygen is not present, a polyglycol's biodegradability is reduced and fluid can accumulate. Its water solubility and its effect on toxicity is under study and because of their compatibility problems, polyglycols are less favored than others; hence their usage as biodegradable fluids is limited.

The technical requirements formulated by German Association of Machinery and Engineering (VDMA) for these three fluids are given in Table 7–3; note that because biodegradability is a new technology, the standards for biodegradable fluids are under close review. The data on commercially available biodegradable fluids is given in Appendix 3, Tables A3–1 to A3–6, which will give users a quick overview of the available fluids as well as examine the available options [40].

MAINTENANCE ASPECTS

The usual maintenance care taken with mineral oils, like leakage and contamination control, is also applicable to biodegradable fluids. Since these fluids are still in the development stage and application is not widespread, a

Table 7–3   Minimum Technical Requirements for Biodegradable Hydraulic Fluids Per VDMA Standard 24568

Type and ISO viscosity grade	HETG 22/32	HETG 46/68	HEES 22/32	HEES 46/68	HEPG 22/32	HEPG 46/68	Test procedure	Remarks
Kinematic viscosity, cSt							DIN 51550 in Conjunction with DIN 51561, 51562, and 51568	
@ 0°C maximum	300/420	780/1400	300/420	780/1400	300/420	780/1400		
40°C	22/32	46/68	22 / 32	46/68	22/32	46/68		
100°C minimum	4.1/5	6.1/7.8	4.1/5	6.1/7.8	4.1/5	6.1/7.8		
Low temperature fluidity after 7 days	*	*	N.A.	N.A.	N.A.	N.A.	ASTM D 2532	
Pour point maximum °C	*	*	–21/–18	–15/–12	–21/–18	–15/–12	DIN ISO 3016	
Flash point minimum °C	165/175	185/195	165/175	185/195	165/175	185/195	DIN ISO 2592	
Insoluble contents %			Below the limit of quantification				DIN ISO 5884	
Water content maximum in mg/kg	1,000	1,000	1,000	1,000	5,000	5,000	DIN 51777 Pt. 1& 2	
Steel corrosion test maximum	0-A	0-A	0-A	0-A	0-A	0-A	DIN 51585	
Copper corrosion test maximum	2-100 A3	2-100 A3	2-1000 A3	2-100 A3	2-100 A3	2-100 A3		
Baader oxidation test							DIN 51759	@ 95°C72 hours for HETG fluid
Increase in viscosity @40°C, %	<20	<20	<20	<20	N.A.	N.A	DIN 51554 Part 3	@ 110°C72 hours for HEES fluid
Aging test: neutralization number increase after 1,000 hours in mg KOH/g	N.A.	N.A.	N.A.	N.A.	2.0	2.0	DIN 51587	
Behavior towards sealing material at °C after 1,000 hours	80	80	80	100	60/80	100	CETOP R81 H Other related standards–DIN 53505, 53521, 53504, and 51381	HNBR for HEES and AU for HEPG fluid are not suitable as standard reference elastomer (SRE).
HNBR–Change in shore A hardness	±10	±10	±10	±10	±10	±10		

Property							Standard	Remarks
FPM–Relative volume change maximum %	−3/+10	−3/+10	−3/+10	−3/+10	−3/+10	−3/+10		The fluid must satisfy at least two SREs, i.e., AU/FPM or AU/NBR or NBR/FPM except for HEES at 100°C (only FPM)
NBR 1–Decrease in elongation maximum %	30	30	30	30	30	30		
AU–Decrease in tensile strength maximum %	30	30	30	30	30	30		
Air release @ 50°C maximum in minutes	7	10	7/10	10	7	10	DIN 51381	Not a guarantee against attack on materials & hydrolytic stability
Foam test mL. Seq. 1	150/0	150/0	150/0	150/0	300/0	300/0	ASTM D 892-89	
Seq. 2	75/0	75/0	75/0	75/0	300/0	300/0		
Seq. 3	150/0	150/0	150/0	150/0	300/0	300/0		
Demulsification @ 54°C in minutes	*	*	*	*	N.A.	N.A.	DIN 51599	This test not applicable to VG 22
FZG gear test–load stage fail	N.A/10	10	N.A/10	10	N.A/10	10	DIN 51354 Part 2	A differentiating evaluation below the stated limiting value is not possible
Mechanical testing in the vane pump ring weight loss maximum in mg	120	120	120	120	120	120	DIN 51389 Part 2	
vane weight loss maximum in mg	30	30	30	30	30	30		
Density @ 15°C kg/m³	*	*	*	*	*	*	DIN 51757	Used for proof of identity for checking incoming goods.
Ash content (oxides) %	*	*	*	*	*	*	DIN EN7	
Neutralization value in mg KOH/g	*	*	*	*	*	*	DIN 51558 Part 1	Dependent on base oils and special additives, specifying general limiting value is not possible

VDMA – German Association of Machinery and Engineering; N.A-Not applicable; *shall be given by the supplier
(Courtesy: The British Fluid Power Association, Extract from *Report BFPA/P 65-1995*)

well-established maintenance procedure has yet to emerge. The fluids' long-term performance characteristics, such as aging and the effect of fluid degradation on performance and components, are still under close study. Hence the standard limits for various performance parameters have not been established. Nevertheless, based on experience and available data, there are guidelines available that have been put into practice. The parameters to be monitored for triglycerides and synthetic esters are

- Water content
- TAN
- Viscosity
- Calcium content

Per the present norm, the limit for water content is less than 0.1 percent [32, 36, 37]; however, a recommendation has been made that it be less than 0.05 percent [33, 37]. As said earlier, water kills fluid life in the shortest time, so it is essential to take all possible measures against water entering the system and to keep the water content as low as possible. A moisture trap should be incorporated into the breather. The precautions against water contamination that have been discussed for mineral oils and phosphate esters are applicable also to biodegradable fluids. Water contamination must be monitored regularly, as it severely affects many performance characteristics. Discard the fluid when the water content exceeds 0.2 percent [37].

When the TAN number of triglycerides reaches a value between 1.5 and 2, the fluid is not serviceable and must be discarded. With synthetic esters, some fluid manufactures permit a TAN value of up to 5 [37], as these fluids have better thermal and oxidation stability than triglycerides. As in other type of fluids, viscosity is monitored regularly, and the allowable change in viscosity for both fluids is ±10 percent. When this limit is exceeded, the fluid must be changed. Another parameter that must be monitored is calcium content of the fluid, which should not exceed five ppm [37]. Increase in calcium content can interfere with the fluid's filterability; water contamination can also increase the calcium in the fluid.

Regular monitoring of water content is also important with polyglycols. Water absorption and the consequent increase in water content can reduce the viscosity of the fluid, which can reduce its load-carrying ability; this can make the fluid corrosive, which may lead to the failure of the pump [38]. Water content must therefore be monitored regularly. The aging properties of polyglycols are closer to those of mineral oils, and hence, as stated earlier, the oil-change period is generally the same as that for mineral oils.

It is always advantageous for users to monitor various parameters at regular intervals and log them. This will help in understanding the trend of fluid degradation. Based on this data, fluid service life can be established; such data would also help to understand the causes of fluid degradation so that preventive measures can be initiated. As the data generated is based on the actual operating conditions of the user's application, the established procedure will be very reliable.

For mobile applications, Foelster and Harms have experimented with a novel on-line monitoring method by means of a surface-shear-oscillation sensor to measure the temperature and viscosity of the fluid [41]. The information obtained is analyzed by Fuzzy logic to assess the fluid's aging; this method gives scope to monitor other parameters like water content and flow rate. Since maintenance parameters and their limits have yet to be established for biodegradable fluids, on-line monitoring will allow users to change fluids confidently and on time, which can prevent the malfunction of hydraulics and possible damage to machine components. An added advantage is that users may be more interested in the application of biodegradable fluids.

CHANGEOVER PROCEDURES

The VDMA guidelines for changeover from mineral oil to biodegradable fluids are given in Table 7–4. As with other fluids, the compatibility of these fluids with hydraulic components such as pumps, valves, filters, seals, hoses, etc., should be closely examined, particularly with respect to operating conditions. Generally, changeover from mineral oil to tryglycerides or synthetic esters is not much of a problem, as their compatibility with mineral oil is good; however, the difference between the two fluids should be clearly understood, and careful consideration should be given to all factors such as pressure, speed, operating temperature, and environmental conditions. Most of the points outlined for changing over from mineral oil to fire resistant fluids given in Chapter 5 are also applicable here. The fluid supplier should also be consulted regarding the suitability of hydraulic hardware and the materials used, as well as for operational guidance.

As natural or synthetic esters are water sensitive, to prevent water from getting into the system, incorporate a moisture trap in the breather. Examine the system carefully and take extra measures to seal it from moisture and water wherever required. In a system that will operate with natural or synthetic esters, it is worthwhile to install a temperature sensor for both the lower and upper limits.

It is important to work closely with the equipment and fluid suppliers, so that their experience can be shared and appropriate precautions taken at the design stage itself. It's important to train all concerned personnel about the properties and application of biodegradable fluids so that they will be fully equipped to handle the new technology. Bear in mind, human error is the one often causes new technologies to fail. Training the people at all levels, therefore, is very important for the successful application of biodegradable fluids.

QUALIFICATIONS FOR BIODEGRADABLE FLUIDS

There are many test methods, as given in Table 7–1 [11, 13, 42 to 46], to assess biodegradability, and most of them are formulated for water-soluble chemicals such as detergents. Hence their application to hydraulic fluids, which are water insoluble, must be done with caution. In other words, many current test procedures are not specific to hydraulic fluids or lubricants. Though the test procedures permit the use of emulsifiers (limited to 100 ppm only) to improve the

Table 7-4 Guidelines for the Application of Biodegradable Hydraulic Fluid

Change of fluid		Temp	Compatibility of the fluid					Measures during and after changeover	
From	To		Seals, plastics, and adhesives	Metals	Filter elements	Paints	Residual volume	Oil-change period	Filter change period
HL HM HV	HETG	−10 to +70°C	Check for plastics and soluble adhesive compounds. For seals, refer to Table 7–2	Avoid lead, tin, zinc in pure form. Alloys of these metals are subject to corrosion with the aged fluid at elevated temperature	Zinc-coated filters are subject to attack	PU-bound zinc dust paint used for interior may not be compatible	2% max	First change after 500 working hours. Second change after 1,000 working hours. Then change after every 2,000 hours. Monitor parameters such as water content, viscosity, and contamination, etc. every 300 hours	Change after the first 50 hours. For further change, establish the period
HL HM HV	HEES	−20 to +80°C	Check for plastics and soluble adhesive compounds. For seals, refer to Table 7–2	Avoid lead, tin, zinc in pure form. Alloys of these metals are subject to corrosion with the aged fluid at elevated temperature	Zinc-coated and paper filter cartridges are subject to attack	PU-bound zinc dust paint used for interior may not be compatible	2% max	First change after 500 working hours. Second change after 2,000 working hours. Then change after every 2,000 hours. Monitor parameters such as water content, viscosity, and contamination, etc.	Change after the first 50 hours. For further change, establish the period

(continues)

HL HM HV	HEPG	−20 to +80°C	Check for plastics and soluble adhesive compounds. For seals, refer to Table 7–2	Avoid lead, tin, and zinc in pure form as well as frictional combination with aluminum. Alloys of these metals are subject to corrosion with the aged fluid at elevated temperature	Zinc-coated and paper filter cartridges are subject to attack	Depends upon the type of glycols used. Check with the supplier	1% max	First change after 500 working hours. Second change after 2,000 working hours. Then change after every 2,000 hours. Monitor parameters such as water content, viscosity, and contamination, etc.	Change after the first 50 hours. For further change, establish the period

(Extract from the British Fluid Power Association *Report P65-1995*, VDMA standard 24569, and Mannesmann Rexroth document *RE 03 145/05.91 and RE 90221/05.93*)

solubility of the substance, this may not make the test conditions effective in actual environment conditions.

As with other test procedures, modeling environmental conditions and taking into account all possible situations is a difficult task, as the parameters are many and varied—water, soil, aerobic and anaerobic conditions, marine life, plants, the many types of aquatic organisms, to name only a few. In the actual environment, microbes are "freelance," whereas in a test they are "captive." This means that in the environment microbes have access to many easily digestible resources, but in the test, they have only the test compound. In the actual environment, therefore, the degree of biodegradation may not be the same as that observed in the test. Hence many test standards are either in the formative stage or under closer study and revision [12]; this has given rise to many test standards, and each method has its own merits and limitations. So, for a specific application, a fluid may have to comply with more than one standard. This also suggests that comparison of one product with another must be done with caution. It should be noted that the test conditions prescribed in most procedures are slightly less favorable for biodegradation. The test temperature used is between 20°C and 25°C, but in tropical countries like India, where the ambient temperature is always higher than that, biodegradation is faster. Fluids that pass through these tests are therefore likely to biodegrade faster in actual environmental conditions.

Considering the large number of test procedures available, it is beyond the scope of this book to detail all of them. Only the test procedures commonly used for hydraulic fluids are outlined here.

Primary Biodegradability Test

As mentioned earlier, the first step in the biodegradation process is primary biodegradation. Hence, the primary biodegradability test of the substance is the prima facie evidence of its capacity to biodegrade. The test procedure used to assess primary biodegradability is the Coordinating European Council (CEC) test procedure, which is more widely recognized and commonly featured in the specification of biodegradable hydraulic fluids.

CEC-L-33-A-93 Test Method This test [44] is intended to assess primary biodegradability and is a test of relative biodegradability, as it is conducted against two-reference oils—a poorly biodegradable white oil (30–40%) and a readily biodegradable synthetic ester, DITA (85–90%). This test was developed originally to evaluate two-stroke outboard engine oils, and it is the only test specific to lubricants. Also, it has widespread industry acceptance. The method is based on the principle that the rate of biodegradation depends upon the quantity of conversion of hydrocarbons (CH_3-CH_2group) into carbon dioxide.

A small quantity of test fluid dissolved in a defined organic test medium such as trichlorotrifluoroethane (or tetrachloroethene) is inoculated with municipal sewage organisms. The mix is incubated along with flasks containing blanks poisoned with mercuric chloride (but without added microorganisms) for 21 days at 25°C in the dark with continuous shaking. After 21 days, the fluid is analyzed by infrared (IR) spectroscopy to estimate the quantity of oil remaining in the fluid; i.e., the concentration of hydrocarbons present (left undigested) is measured. This is done by measuring the maximum absorption of the CH_3-CH_2 band at 2930 cm^{-1}. The loss of product is measured with respect to the poisoned flasks. From this, biodegradability is calculated as the percentage difference in the residual oil content between the poisoned control flasks and the respective test flasks. Comparison with the values obtained for two calibration flasks gives a measure of the quality of the test performed.

$$\% \text{ Biodegradability} = [(P - T)/P] \times 100$$

Where P = Mean value of residual oil content of poisoned flask in %
    T = Mean value of residual oil content of test flasks in %
A minimum of 80% is required for the fluid to qualify as biodegradable, per Germany's Blue Angel criteria.

Ultimate Biodegradability Tests

As explained in fig. 7–1, passing the CEC test means, the fluid has reached only the first end point of biodegradation. It is possible that a fluid that gets to the first end point of the biodegradation process may not get to the second end point—ultimate biodegradation. In other words, after primary biodegradation, the substance can remain a recalcitrant product of different physical and chemical properties. Hence the fluid must be assessed for ultimate biodegradability by one of the ultimate biodegradability tests. A test often used as a complement to the CEC test is Organization for Economic Cooperation and Development (OECD) 301-B, CO_2 evolution test (modified Sturm test). Apart from the test's simplicity, its suitability to evaluate both water-soluble as well as water-insoluble substances (such as lubricants) and the lower concentration requirements have made this test more acceptable for assessing the ultimate biodegradability of hydraulic fluids.

OECD 301-B Test This test is based on the CO_2 evolved during biological decomposition of the test material (lubricant)—biodegradability of the test fluid is measured by collecting and quantifying the CO_2 produced when the test fluid is exposed to microorganisms under the prescribed aerobic aquatic conditions. The CO_2 evolved is the product of aerobic microbial metabolism of carbon present in the test fluid (it may be recalled that lubricant is a hydrocarbon substance), and hence it is a direct measure of a fluid's ultimate biodegradation. The CO_2 thus evolved is compared with the theoretical quantity of CO_2 that would be generated if all the carbon present in the test fluid were converted to CO_2. Biodegradability is expressed as a percentage of the theoretical CO_2 produced.

Per this procedure [45], the test fluid of a known concentration placed in a defined inoculated mineral medium is incubated for 28 days. The inoculum used may be derived from a variety of sources—activated sludge, sewage effluents, surface waters, soil, or a mixture of these. During the incubation period, the test fluid is aerated by air free of carbon dioxide at a controlled rate in the dark or in diffused light at 22° C and also agitated by a magnetic stirrer. As the test fluid breaks down due to the metabolic action of microorganisms, carbon dioxide is produced, which is trapped in a solution of barium or sodium hydroxide. The quantity of carbon trapped is determined either by a carbon analyzer or by titrating the solution. The percentage of biodegradation is calculated from the formula:

$$\% \text{ Degradation} = \frac{\text{carbon dioxide produced in mg}}{(\text{theoretical } CO_2) \times (\text{test substance added in mg})} \times 100$$

The results indicate the final degradation of the fluid into carbon dioxide and water. A minimum of 70% biodegradability is required for the fluid to qualify as readily biodegradable.

ASTM D 5864 Method The ASTM D 5864-95 standard [13] is another test procedure that measures the carbon dioxide produced under controlled aerobic aquatic conditions. It is similar to the OECD test but is specifically meant for lubricants, and the test is conducted in the dark to prevent photodegradation of the test substance and growth of photosynthetic bacteria and algae. Another difference is that the option to test the fluid with pre-adaptation or pre-acclimation of the inoculum is available. Pre-adaptation means that, prior to the initiation of the test, the inoculum is incubated with the test substance under conditions similar to the test conditions, which helps the microorganisms adjust their metabolism to the test substance. The advantage is that the test can be more easily repeated, as the inoculum used is already accustomed to the test substance and the test environment.

It's important to note that in all the test methods, as the inoculum is not standardized, it could cause statistical dispersion in the test results; pre-adaptation of the inoculum may mitigate this problem. In the OECD test only pre-conditioning of the inoculum is allowed, i.e., aerating the activated sludge (in a mineral medium) or secondary effluent for five to seven days at the test temperature with a view to improving the precision of the test method by reducing the blank values. A pre-adapted inoculum is not allowed in the OECD test.

EPA Shake Flask Test This test method [42, 43] is also similar in principle to both the OECD and ASTM procedures, but differs in the laboratory equipment used. The test is conducted in a specially equipped two-liter Erlenmeyer flask (fig. 7–2). A specially made reservoir containing barium hydroxide—$Ba(OH)_2$— solution is suspended in the test flask over a one-liter culture medium. The reservoir is made by attaching a 50 mL heavy duty centrifuge tube to a glass tube of 9 mm O.D. × 3 mm I.D. by means of three glass supporting rods. The constriction placed just above the 10 mL mark in the centrifuge tube facilitates the transfer of the flask to and from the shaker flask without spilling the $Ba(OH)_2$ into the medium, and the opening in the centrifuge tube is large enough to permit CO_2 to diffuse into the $Ba(OH)_2$. For periodic removal and addition of the base from the center well, a 2 mm O.D. polypropylene capillary tube attached to one end of a 10 mL disposable syringe is inserted through the 9 mm O.D. glass tube into the $Ba(OH)_2$ reservoir. Sparging, medium sampling, and flask additions are done through two glass tubes.

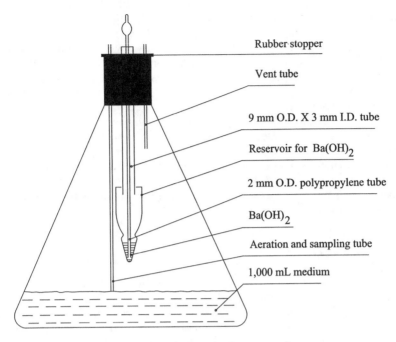

Rubber stopper

Vent tube

9 mm O.D. X 3 mm I.D. tube

Reservoir for Ba(OH)$_2$

2 mm O.D. polypropylene tube

Ba(OH)$_2$

Aeration and sampling tube

1,000 mL medium

FIGURE 7–2   EPA Shake Flask Test Apparatus

As with the ASTM method, this test is also conducted in the dark, but the inoculum used is only from soil and sewage microorganisms. A major difference from the other two methods is that as a standard procedure, the inoculum used is pre-adapted to the test substance. Throughout the test period, the test flasks are agitated by a gyrotary shaker at approximately 125 rpm.

A common feature of all these test methods is that in addition to the fluid under review, one or two readily biodegradable reference compounds are also tested in parallel as part of the test runs. This is done to assess the microbial activity of the test medium, and in simple terms it provides an internal check for the procedure. For example, the suggested reference fluid for the ASTM procedure, which is specific to lubricants, is low erucic acid rapeseed oil (LEAR). Per the ASTM procedure, the test is invalid if the biodegradability of the reference fluid is less than 60 percent, indicating that an error or irregularity crept in during the conduct of test. In such cases the test has to be rerun using a fresh inoculum. In the EPA method the reference compound should yield a final CO_2 evolution value in the range of 80 to 100 percent of theoretical CO_2; otherwise the test results must be considered invalid and the test rerun.

The preparation of the test substance, inoculum, pre-adaptation of the inoculum, and test apparatus are quite elaborate. Also, a great number of flasks and samples are used to conduct the tests and data collections, which requires a certain level of skill and precision. For details, readers may refer to the relevant standard.

ECOTOXICITY

A fluid qualified as biodegradable is not necessarily environmentally acceptable. Nonbiodegradable components of a biodegradable fluid in an aquatic environment, for example, can produce toxic substances harmful to fish or other aquatic species. In other words, the fluid can be ecotoxic. Evaluation of ecotoxicity is, however, very complex, as there are many types of organisms and hence the effect of a fluid on different plants and animal can vary considerably. One test on a single species may not adequately evaluate a substance's ecotoxicity. However, one widely used test to evaluate ecotoxicity is to determine the lethal concentration (LC_{50}), which is defined as that concentration of a test substance in water needed to kill 50 percent of the test batch of fish within a particular period of exposure. This parameter is widely recognized by regulatory bodies worldwide to assess the toxicity of any material.

OECD 203 Test

The OECD 203 [46] test is also called the "fish, acute toxicity test." Fishes are exposed to the test substance for a period of 96 hours under controlled conditions. For example, the oxygen concentration should not be less than 60 percent of the air saturation value, there is no feeding or disturbance of the fish during the test period, and 10 to 12 hours of light (photo period) in a day are allowed. Mortalities are recorded at 24, 48, 72, and 96 hours, and the concentration that kills 50 percent of the test fishes (LC_{50}) is determined by statistical procedure. The generally accepted value of LC_{50} for the fluid to qualify as toxic is that the concentration should be greater than 1,000 ppm, i.e., a concentration greater than 1,000 ppm of the fluid in an aqueous solution is needed to achieve a 50 percent mortality rate of the fish [47, 48]. There are many recommended test species per the test procedure, and the test temperature depends upon the species used. The commonly used test fish is the rainbow trout.

Water Endangering Number

A fluid must pass through another ecotoxicity parameter—the "water endangering number" (Table 7-5). The water endangering number (WEN) is a combined measure of toxicity to mammals, bacteria, and fish. Biodegradable fluids must come under the classification WGK 0-1 (Wasser Gefahrdungs Klasse), if they are to qualify against ecotoxicity. The WGK number for mineral oils, for example, is 2. This is a comprehensive and effective instrument to assess the toxicity of a lubricant in an aquatic environment.

ECOLABELING

It can be seen that for a biodegradable hydraulic fluid to qualify as environmentally acceptable, it must pass a variety of tests. Ecolabeling identifies what is

Table 7–5 The German Water Hazard Classification System–Calculation Procedure [10]

Acute oral mammalian toxicity (AOMT)
Measured as an LD_{50}(lethal dose, 50% death rate) for laboratory animals. The LD_{50} is used to determine a water endangering number (WEN) using the following scale:

LD_{50} in mg/kg	WEN (AOMT)
<25	7
25–200	5
200–2,000	3
>2,000	1

Acute bacterial toxicity (ABT)
Measured as a "no-effect" concentration (NEC). The WEN (ABT) is determined as follows:
WEN (ABT) = $-\log$ (NEC in ppm/10^6 ppm)

Acute fish toxicity (AFT)
Measured as an NEC. The WEN (AFT) is determined as follows:
WEN (AFT) = $-\log$ (NEC in ppm/10^6ppm)

Water endangering number
The overall WEN is calculated using the expression:
Overall WEN = [WEN (AOMT) + WEN (ABT) + WEN (AFT)]/3

WGK Number
The water hazard classification is then determined from the overall WEN:

WEN	WGK Number	Classification
0–1.9	0	Not hazardous to water
2–3.9	1	Slightly hazardous to water
4–5.9	2	Moderately hazardous to water
>6	3	Highly hazardous to water

environmentally acceptable. Ecolabeling systems exist in many countries—the U.S. (Green Seal and Green Cross), Germany (Blue Angel), Canada (Maple Leaf), Scandinavia (White Swan), Japan (Ecomark), India (Ecomark). Among them, Germany's Blue Angel specification (Table 7–6) is the pioneer and is commonly quoted.

Germany introduced the Blue Angel as early as 1977 [17]; it was the first country to use an ecolabeling program and hence it has been a model and inspiration to other countries. It is not easy to comply with the stringent Blue Angel criteria. However, it is encouraging to note that fluids with the Blue Angel label are commercially available [6, 36]. So far, the Blue Angel label has been awarded to 27 hydraulic fluids. In the absence of appropriate government regulations, ecolabeling is an assurance that fluids complying with the ecolabeling scheme are reliable, environmentally friendly products. An ecolabeling system common to all members of the European Union is under way.

Because biodegradable fluid is a new technology, the market is charged with many terminologies that often confuse consumers and sometimes deliberately mislead them. Environmentally friendly, environmentally compatible, environment sparing, biodegradable, ecotoxic, ecologically acceptable, green fluids, biofluids, etc., are just a few of the terms that are tossed around.

Table 7–6 Blue Angel Specifications for Environmentally Acceptable Fluids

I Prohibitions
- No components of WGK 3
- No chlorine, no nitrides
- No components that are listed in the GefStoff V
- Calcium content max.0.1

II Ecological tests
a) Maximum additive content 7%   5% potential biodegradability
(Zahn-Wallens test, minimum 20%)
2% nonbiodegradability but
no bioaccumulaion and no bacterial
toxicity
b) Minimum 80% biodegradability according to CEC-L-33-A-93 test and BODIS test
or
Minimum 70% biodegradability according to AFNOR, MITI, STURM OECD test
c) Ecotox testing of the complete formulation against:
- fish, daphnia, and plant
- plus bacteria-restrained test
- algae
d) Limited accumulation potential

III Manufacturers declaration
- Detailed formulation
- Field of application/conformity with technical standard (VDMA 24568)
- Inform customers of correct disposal method and potential reuse

Government guidelines are obviously necessary. A notable one is the U.K.'s Department of Trade and Industry [3] guidelines to help advertisers.

Do: Make sure all the claims can be sustained.
Make clear the basis of your claim.
Make relevant or qualified claims as distinct from absolute claims.
Don't: Try to blind the customer with science.
Mislead by omission.

FUTURE OUTLOOK

At present it seems that with the exceptions of Germany and Switzerland, where it is already mandatory to use biodegradable fluids in environmentally sensitive applications such as forestry and mining, other countries are only beginning to frame regulations for the use of biodegradable fluids. In the U.S., for instance, per EPA regulations, no visible oil sheen can be evident downstream from equipment located in or near waterways [48]. Even an oil concentration of 10 to 15 ppm can create a visible oil sheen, and hence the use of mineral oils in such areas is virtually prohibited. In the U.S., the army has tested a number of commercially available biodegradable hydraulic fluids and found that most of them are close to its specifications. The army has also embarked on field trials and found performance satisfactory per the interim results [30, 47, 49]. New initiatives taken by the government at the state and federal level in the U.S. to expand the application of biodegradable fluids through legislation as well as price support are also promising [51]. Rapeseed-based hydraulic oils are already used for the operation of lock gate operating machinery, excavators, cranes, and dredges in hydropower plants, and based on this experience, Army Corps of Engineers Installations are intended to expand the use of these oils [50]. In Germany, Kutter and Feldmann have established the suitability of synthetic esters for operating low-pressure vane type hydraulic motors used in ship deck machinery [52]. As mentioned earlier, biodegradable fluids are already in use for machine tool hydraulics and metal cutting operations [31]. A range of biodegradable fluids is available in many countries. In India, the major lubricant company, Indian Oil Corporation, makes more than a dozen biodegradable fluids [8]. And the last decade has witnessed extensive research work on the evaluation of various fluid properties, test methods, operational problems, monitoring, and application of the fluids at various research centers [14 to 26, 30, 35, 41, 51 to 63].

At this stage, it is appropriate to assume that the process of evolution is on, that the current decade will see a rational and more orderly use of biodegradable fluids, and that government regulations will catalyze this process. Ongoing development work may also improve the technological capabilities of these fluids and help expand their usage. Cooperative efforts of fluid suppliers, system designers, equipment builders, and plant personnel are required so that a closed-loop feedback system can be established. The data collected in the field as well as in the labs should be exchanged and collated for better understanding of both the short-term and long-term behavior of fluids under different working conditions, which will overcome many of the uncertainties encountered in the present usage of these fluids.

At present, the option to recycle is unavailable to users of biodegradable fluids, and as such users may think that the additional disposal cost may be an another burden to bear. This problem must be resolved so as to improve the competitiveness of biodegradable fluids. As the composition of these fluids is made up of environmentally compatible substances, this may not pose much of a problem.

It is indeed interesting to note that Dowson's extensive research on the history of tribology [64] indicates that our civilization started with the use of natural oils and fats as lubricants. But industrial civilization replaced these natural products with man-made ones, bringing a host of environmental problems. In this new millennium, we are reconsidering the use of vegetable oils as lubricants for a variety of applications, as postulated by Padavich and Honary [27]. Have we come full circle?

Chapter 8

MAINTENANCE OF HYDRAULIC FLUIDS

The full benefit of the best fluid and equipment cannot be harnessed if the fluid is not properly cared for. The majority of hydraulic equipment breakdowns are fluid-related. In addition, the sophistication of modern hydraulic equipment requires clean fluid, and because the cost of lubricants and hydraulic fluids is increasing, it is imperative to pursue a sound fluid maintenance schedule. Table 8–1 shows the typical operating clearance of various machine elements, which emphasizes the importance of keeping fluid clean. Yet despite the significant bearing maintenance has on the operating costs of a plant, it is often a neglected engineering discipline. Hence a detailed account on the maintenance aspects of hydraulic fluids is given in this chapter.

MAINTENANCE OF MINERAL HYDRAULIC OILS

The care of hydraulic oils starts before they are put into the system; in fact, it begins at the storage stage itself. During storage oil barrels should be kept in a horizontal position to prevent water or other contaminants from collecting near the bungs. This practice also ensures that the gasket inside the bung is kept wet so that it can be more effective as an airtight seal. The ingress of airborne moisture or the formation of rust in the drum in the airspace above the fluid can also contaminate oil during storage. The contamination level of hydraulic oil as supplied by manufactures is at best closer to ISO class 20/18/15 (Table 8–2) or NAS 9–10 (Table 8–3), which is not acceptable to many hydraulic equipment manufacturers. For example, per the NAS 9 level, the number of 15 to 25 μm nominal particle size per 100 mL of oil is around 22,800, which electro hydraulic servo valves, precision hydraulic mechanisms, and high-pressure pumps cannot tolerate, they may require a cleanliness level of ISO class 17/15/12 (NAS 6). Approximate equilavalents between ISO and NAS contamination codes are given in Table 8–4. It is necessary to keep the contamination level of hydraulic systems as given in Table 8–5 or as recommended by equipment manufacturers. It should be borne in mind that specifying the contamination level/ particle count per a standard is only a convenient form that is easy to remember and follow. Usually the count level between classes varies by a factor of two. With

Table 8–1 Typical Operating Clearances for Various
Machine Elements

Component	Clearance μm
Rolling element bearings	0.1–1
Journal bearings	0.5–100
Hydrostatic bearings	1–25
Dynamic seal	0.05–0.5
Pump, Gear	
Tooth to side plate	0.5–5
Tooth top to case	0.5–5
Pump, Vane	
Vane sides	5–13
Vane tip	0.5–1
Pump, Piston	
Piston to bore	5–40
Valve plate to cylinder	0.5–5
Servo Valves	
Orifice	130–450
Spool to sleeve	1–4
Flapper wall	18–63
Actuators	50–250

such variations, it is advisable to run the system at well below the recommended particle count limits for trouble-free operation.

It is often important to use an auxiliary filtering unit capable of filtering 10 μm nominal particle size to pump the fluid from the barrel to the system tank. Though some hydraulic equipment may be able to tolerate the ISO 20/18/15 contamination level, prefiltration to ISO 18/16/13 level is essential; otherwise the filter will clog within a few hours of operation, which can damage the servo valve if the system is not built with a safety device. This also brings us to the fact that oil charged should be prefiltered at least to the level of cleanliness the filters are expected to maintain.

Handling Superclean Fluids

Systems designed with very fine tolerances and micro filtering devices call for superclean fluids. These requirements were traditionally found in aerospace and military applications. However, they are now also required for ultraprecision applications and testing devices. What is a superclean fluid? As of now, it appears that there is no accepted definition for superclean fluids. In other words, the contamination level criteria for superclean hydraulic fluids has yet to be defined. However, it seems that the fluids meeting the cleanliness level of NAS 3 or better can be called superclean. It is obvious that these fluids need greater care in handling, from the storing to the filling stage. These fluids are stored only in clean rooms to prevent airborne contaminants from getting into them. All pipelines are subjected to ultrasonic cleaning before they are fitted. Before the system is put into operation, it is flushed with the same oil under pressure, say, 10 bar in a clean room with a suitable bypass wherever required,

Table 8–2 Allocation of ISO Scale Numbers

Number of particles per mL		
More than	Less than	Scale number
2,500,000	–	>28
1,300,000	2,500,000	28
640,000	1,300,000	27
320,000	640,000	26
160,000	320,000	25
80,000	160,000	24
40,000	80,000	23
20,000	40,000	22
10,000	20,000	21
5,000	10,000	20
2,500	5,000	19
1,300	2,500	18
640	1,300	17
320	640	16
160	320	15
80	160	14
40	80	13
20	40	12
10	20	11
5	10	10
2.5	5	9
1.3	2.5	8
0.64	1.3	7
0.32	0.64	6
0.16	0.32	5
0.08	0.16	4
0.04	0.08	3
0.02	0.04	2
0.01	0.02	1
0.00	0.01	0

and the contamination level of the flushing oil is periodically monitored. The flushing is continued until the contamination level of the flushing oil reaches at least two grades better than the system requirement. Before the fluid is charged into the system, it is filtered off-line till its contamination level is at least two grades better than the system requirement. It's also important that all hydraulic components that go into the system go through the established cleaning process before the system is assembled. When the system is operating, the contamination level of superclean fluids are better controlled by periodic off-line filtering rather than in-line filtration devices alone.

Sensory Inspection

The parameters required for periodic monitoring are only a few and so are easy to implement. The first step is sensory inspection. In this age of computers, we tend to dismiss sensory inspection. But an experienced user will have the skill to

Table 8–3 NAS Classification of Contamination Level

Size range μm	Class/maximum number of particles per 100 mL													
	00	0	1	2	3	4	5	6	7	8	9	10	11	12
5–15	125	250	500	1k	2k	4k	8k	16k	32k	64k	128k	256k	512k	1,024k
15–25	22	44	89	178	356	712	1,425	2,850	5.7k	11.4k	22.8k	45.6k	91.2k	182.4k
25–50	4	8	16	32	63	126	253	506	1,012	2,025	4,050	8,100	16.2k	32.4k
50–100	1	2	3	6	11	22	45	90	180	360	720	1,440	2,880	5,760
>100	0	0	1	1	2	4	8	16	32	64	128	256	512	1,024

$K = 10^3$

Table 8–4 Approximate Equivalents of
ISO-NAS Contamination Codes

ISO 4406-1999	NAS 1638
13/11/8	2
14/12/9	3
15/13/10	4
16/14/9	–
16/14/11	5
17/15/9	–
17/15/10	–
17/15/12	6
18/16/10	–
18/16/11	–
18/16/13	7
19/17/11	–
19/17/14	8
20/18/12	–
20/18/13	–
20/18/15	9
21/19/13	–
21/19/16	10
22/20/13	–
22/20/17	11
23/21/14	–
23/21/18	12
24/22/15	–
25/23/17	–

(From BFPA *Report P82-1999*)
Courtesy: The British Fluid Power Association

recognize the condition of the oil, and from that, ascertain what type of tests need to be done. Virgin hydraulic oils are bright, clear, and odor-free. A hazy or cloudy appearance indicates the presence of water in the oil. Even 75 to 100 ppm of water content will give a hazy appearance. As the water content increases in the oil, the appearance changes from hazy to cloudy. The color of the oil also gives a rough indication of the level of contamination. Virgin oil kept in a clean bottle as a reference can show the degree of contrast between contaminated and clean oil.

Similarly, dark oil is an indication that the oil is oxidized; when darkness is combined with strong odor, the oil's oxidation is almost confirmed. A burnt odor suggests that the oil is affected by hot spots.

Collecting Oil Samples

The maintenance schedule starts with collection of sample oil. The sample collected for inspection or analysis should be representative of the bulk oil in the hydraulic tank. It is important that the samples are taken at the right place, under the right conditions, following the correct procedures, and using the proper sample containers; otherwise variations in test results can occur

Table 8–5   Suggested Contamination Levels for Hydraulic Systems

ISO code	*NAS class	Type of system
15/13/10	4	Very high reliability required. Aerospace and laboratory applications.
17/15/12	6	High-performance servo and high pressure systems. Aerospace and machine tool applications.
18/16/13	7	Proportional control valves and general machinery. High quality and reliability.
20/18/15	9	Medium pressure and medium capacity. Mobile hydraulics and general mechanical engineering.
21/19/16	10	Low-pressure systems, construction equipment, heavy-duty industrial systems, and mobile heavy equipment.

*Approximately equivalent to ISO code

depending upon the way samples are collected [1–6] There are a few ground rules for collecting samples:

- The sampling bottle should be clean and dry.
- Do not collect any stagnated oil.
- Collect the oil when the system is running under equilibrium condition.
- The sample tap should be purged to prevent stagnant fluid from getting into the sampling bottle.
- Take the sample from the same point in the system each time, preferably near pump inlet or return lines.
- If makeup oil is added, allow several hours for thorough mixing before the sample is taken.
- Label the sample without fail.
- If the oil has to be sent to lab for analysis, do so immediately.
- Do not forget to shake the sample thoroughly before conducting any test. This brings the sample oil to a condition similar to running conditions.

The sample oil is used mainly to monitor four important parameters: water content, TAN, viscosity, and contamination level. Techniques such as X-ray diffraction, infrared, ferrography, and a variety of spectroscopic methods are also used to identify the metals present in used oils so that wear of any critical components can be detected [7, 8, 9]. Conducting these analyses in-house is not a common industrial practice; usually specialized oil analysis labs do them. Also, many industrial systems don't require these tests. Only the tests that are routinely used in industries are dealt with here.

Water Contamination

Water contamination due to leaking water from the oil cooler, or due to the moisture condensed from the atmosphere and cutting fluid mixing, are

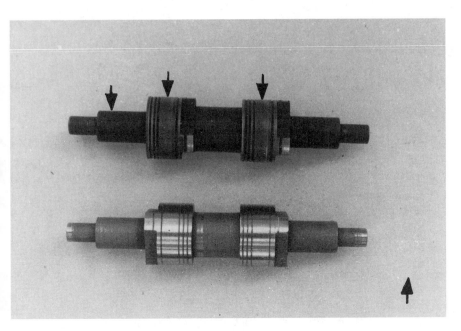

FIGURE 8-1 Corrosion on the Spool Due to Water Contamination (Bottom spool shown in good condition for comparison) (Courtesy: Yuken India Limited, Bangalore)

common problems. Violent churning and dispersion in the pumps and valves can break the water into tiny droplets and help the contaminated water to form an emulsion. The emulsion joins with the other deposits and forms sludge and a sticky substance that impair the performance of pumps, valves, and cylinders. The serious nature of this problem can be appreciated by the fact that in the hydraulic system of a large injection molding machine, frequent burning of solenoids was traced to the valve sticking due to water contamination (unchecked leakage from the cooler) and the consequent oil degradation. Since corrective action was not taken at the right time, the fluid had to be discarded and thorough flushing had to be carried out, which cost a lot of money.

The promotion of corrosion-accelerated bearing fatigue, loss of oil film, silting, additive precipitation or depletion, and acidity increase (oxidation) are a few more problems related to water contamination. Water in concentrations of even 50 to 100 ppm can shorten the life of bearings by as much as 50 percent. In hydraulic systems operating in humid environments, in coastal areas, for example, moisture ingress and the consequent water condensation inside the reservoir is a perennial problem. This problem is also seen where grinding machines are located, as cutting fluids generate a large amount of mist. Any unchecked continuous moisture ingress into the reservoir and the consequent mixing of water with oil can make hydraulic components vulnerable to rust. The rust and corrosive wear on a solenoid valve spool and a vane pump side plate due to water contamination are shown in fig. 8–1 and fig. 8–2, respectively. (In fig. 8–1 the spool at the bottom is in good condition, shown for comparison.)

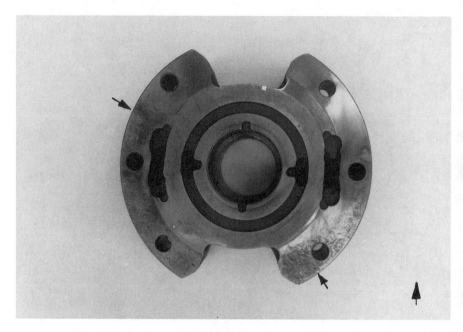

FIGURE 8–2 Corrosion and Corrosive Wear on the Vane Pump Side Plate (Courtesy: Yuken India Limited, Bangalore)

Providing moisture traps in the breather and periodic off-line centrifuging of the oil are some of the steps to be followed to keep oil free of water contamination. In case the system has to operate in hostile conditions, such as dusty or humid environments, an effective solution is to seal the reservoir fully and provide a breather bag, called a hydraulung bag or a flexible separator, in the airspace inside the reservoir. The bag expands or contracts as the oil level changes inside the reservoir. The severity of the operating conditions should be assessed carefully, and appropriate devices incorporated to check water contamination. In one application, the moisture condensation was so great that the user was advised to drain off the water every day. Obviously, this is not a practical solution; it is given only to illustrate that water contamination is not a minor problem. Appropriate preventive measures and periodic checks of water contamination levels, coupled with suitable remedial action, will significantly lower operating costs.

The litmus test for the presence of water in oil is the crackle test. It is a simple, on-the-spot qualitative test. When the contaminated oil is heated in a shallow pan, if audible crackling or popping occurs, it is clear evidence of water contamination. Once the crackle test establishes water contamination of the oil, quantitative tests can be conducted by the distillation method per the ASTM D 95 test procedure (Chapter 3). If the water contamination is more than 0.1%, the oil should be clarified by centrifuging. Centrifuging will remove only free water, not dissolved water. Heating the oil to around 60°C will help the dissolved water to evaporate. Centrifuging will also remove solid contaminants.

Portable automatic dewatering units fitted with a heater and vacuum pump are also available, which can be operated off-line even when the system is under operation [10]. Periodic check and prevention of water ingress into the system should become an essential part of the maintenance schedule.

Total Acid Number (TAN)

This is another parameter to be monitored periodically. In normal service, oil degradation is gradual; accelerated degradation is an indication of either severe service conditions or abnormal operational conditions. As explained earlier, TAN value is used to determine the level of oil degradation and helps in deciding whether to discard the oil or not. Hence the measurement of TAN is informative and useful. The ASTM D 974 (Chapter 3) method to find TAN is simple and inexpensive. But instead of changing the oil based on the TAN value, many users still follow the OEM recommendation to "change the oil every xxx hours." The so-called recommended hours generally include a large margin-of-error by the OEM and can never take into account all the user's specific service conditions and maintenance practices. Periodic check and logging of the TAN value will help the user understand the trend of degradation as well as establish the life of the oil for the given operating conditions. Where large systems are installed or there is a battery of hydraulically operated machines, this practice can bring sizable savings in the oil bill. Because degraded oil can cause the system to malfunction, as explained in Chapter 2, it is essential to monitor TAN and change the oil at the right time.

Because of the presence of certain additives, particularly in zinc-based oils, the virgin oil may show an initial acidity value of as high as 1 to 1.5; often this is called a pseudo TAN value. It is in no way harmful. Because of the presence of oxidation inhibitor in the oil, there is a long induction period (fig. 8–3) before the TAN value reaches 1. But from then on the increase is sharp and quickly reaches the value of 2, the limit for discarding the oil. This is due to the fact that during the induction period, the oxidation inhibitor is consumed and TAN rise is controlled. Once the oxidation inhibitor is depleted, oxidation proceeds rapidly, raising the TAN value to a high level very quickly. Hence when the TAN value reaches 1, it can be taken as a signal to prepare for the changing of the oil. One effective method to clean the oil of oxidation and decomposition products is with electrostatic liquid cleaners [11].

Viscosity

The influence of viscosity on the performance of hydraulic systems has been discussed in detail, and the importance of monitoring the viscosity of the oil periodically needs no emphasis. As the level of oxidation increases, the viscosity also increases markedly. A 15% to 20% increase in viscosity from the original value (at 40°C) can be considered an indication that the oil has reached the end of its life. Reduction in oil viscosity is rare; if it happens, low-viscosity oil mixing with hydraulic oil is likely to be the probable cause. Viscosity can be checked even with a portable viscometer in situ (fig. 8–4).

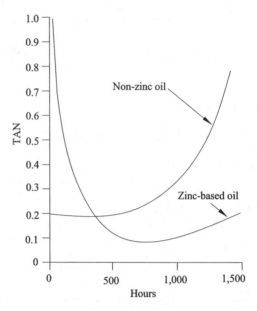

FIGURE 8–3 Typical Pattern of Change in TAN by Oxidation

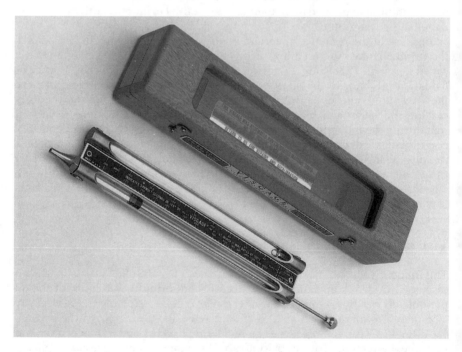

FIGURE 8–4 Portable Viscometer (Courtesy: Stanhope-Seta Limited, U.K.)

Magnetic Traps

Wear particles can accumulate in hydraulic systems and must be flushed out periodically. It is worthwhile to provide a magnetic plug/filter to flush out ferrous contaminants. It should be placed so that it is easily accessible for frequent cleaning; otherwise oil surge in the tank may dislodge the particles from the magnets and circulate them back into the system. Incidentally, the use of magnetic plugs also relieves the load on the filter and increases the filter-change interval; they can also prevent electromagnetic interference when electro hydraulic servo valves are used.

Contamination Control/Particle Counting

The life of the system components depends to a large extent upon how free of contaminants the fluid is. Hence, contaminant control is an important component of maintenance. In fact, more than 50 percent of hydraulic failures are attributed to contamination of the hydraulic fluid [12].Valve lock is often caused by contaminants [12, 13]. Valve lock is caused not just by hard wear particles locked in the clearance between the spool and the sleeve but also by the presence of oxidation and decomposition products. Many hydraulic failures that are traced to poor contamination control confirms this point. Wear on pumps, motors, cylinder rods, and valve seats; erosion of flow-control and valve orifices; and sticking spool valves are common problems caused by contamination. Many equipment manufacturers recommend doing particle counts periodically so that the system operates per the criteria defined by equipment manufacturers. The criteria set by manufacturers depends upon the system design—system pressure and duty cycle, the type of fluid used, life expectancy, and environmental conditions.

Contaminant particles come in many sizes and shapes, depending upon the type of wear that takes place in the system, e.g., corrosive, fatigue, cavitation, abrasive, etc., and the degree of degradation of the oil. The operating environment is another major factor contributing to oil contamination. For instance, in an operating environment where there are many grinding machines around the hydraulically operated machines, grinding dust and moisture are major sources of contaminants; similarly, hydraulic presses used for sintering and machines located near heat treatment shops or foundries are prone to perpetual ingress of metal or sand dust. In such operating environments, better attention must be paid to sealing the hydraulic oil reservoir as well as care of the oil. Providing magnetic traps in the reservoir can also considerably lessen this problem. In severe cases, incorporating a breather bag in the reservoir can provide trouble-free operation. Here again the severity of service conditions must be assessed carefully and suitable devices incorporated into the system.

Yet these simple steps are often breached in practice. In the hydraulic system of a grinding machine, for example, a large amount of grinding dust had accumulated inside the oil reservoir, causing frequent breakdowns. The reason: A

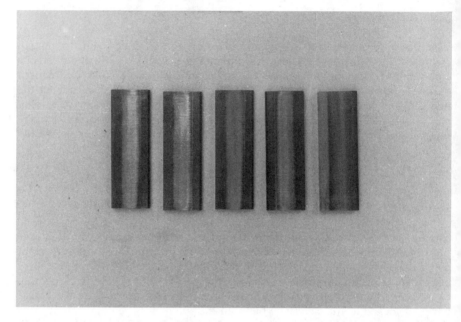

FIGURE 8–5 Scoring Marks on Vane Tips Due to Contamination (Courtesy: Yuken India Limited, Bangalore)

badly sealed reservoir and the air breather provided easy entry for grinding dust into the reservoir! In another case, an oil reservoir received from the supplier was heavily contaminated with cotton swabs and debris; perhaps no user would anticipate such poor practice from the OEM. In another incident, in a precision surface grinding machine of highly reputed make, the table drive cylinder frequently seized due to severe contamination of the oil. The problem was traced to the lack of a filter in the system. Incorporating a filter is basic, but the manufacturer did not recognize that! The correct filter size and its location are important to contamination control [10, 14]. Such instances suggest that in a failure analysis, nothing should be taken for granted; suspect everything and eliminate the causes one by one. Prepare your own checklist for commissioning the hydraulic system to ensure it is charged with oil that meets the desired level of cleanliness.

As mentioned earlier, contamination in the fluid is a major cause of component failure. Some of the contamination-related failures explained below drive home this point. The clearance between the vane and the slot in the vane pump is generally five to eight microns, and hence even fine particles in the oil can cause damaging frosting or scoring marks in the vane as well as in the slot,which require the replacement of the pump (fig. 8–5). When the contaminant particle size is larger than the oil film formed between the vane and the slot, it acts like lapping powder and abrades the surface. Other contamination-related damages are shown in fig. 8–6 and fig. 8–7. Due to the abrasive action of contaminant

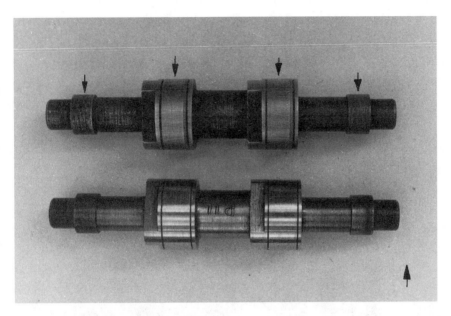

FIGURE 8–6 Abrasion on Spool Due to Contaminations (Bottom spool shown in good condition for comparison) (Courtesy: Yuken India Limited, Bangalore)

FIGURE 8–7 Abraded Relief Valve Plunger Due to Contamination (Courtesy: Yuken India Limited, Bangalore)

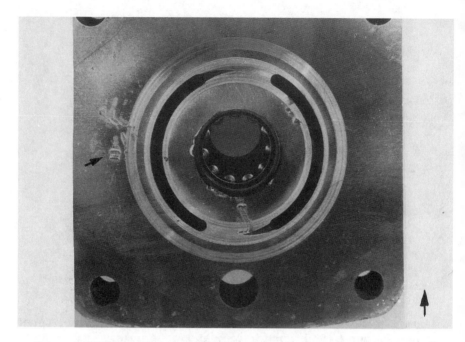

FIGURE 8–8 The Valve Plate Badly Gouged by a Large Piece of Foreign Material (Courtesy: Eaton Corporation)

particles, all the sliding surfaces of the solenoid valve spool (the spool at the bottom is in good condition and shown for comparison) and the relief valve plunger were abraded very badly. Roughened ground surface and the wear on the spool and plunger can be seen in the figure. As a result, fine radial clearance kept between the spool and body, normally 4–12 μm, had increased, and the valves had to be replaced. Another example of component failure due to neglected contaminant control is shown in fig. 8–8, where a valve plate is badly gouged by a large contaminant particle. Besides the replacement cost of the components, in all these case the cost associated with breakdown badly affects the cost of production.

Apart from charging the system with clean oil, proper sizing and location of the filter and its maintenance, cleaner assembly practices, sealing the system against air entry, and control of particle ingression with air breathers or by providing a breather bag inside the reservoir are some of the measures that will control the level of contaminants in the oil. Comprehensive and detailed practical guidance to contamination control is given in the "Vickers Guide to Systemic Contamination Control" [15].

Quantifying Contamination Level Contaminated particles in the oil are quantified by a particle counting technique that determines the number of particles, their size, and distribution; this technique is specified per ISO code (Table 8–2). The ISO code is the accepted system for specifying the cleanliness of hydraulic fluids and systems. The usual method of particle counting is done generally with a microscope or an automatic particle counter (APC). A portable contamination kit [16] normally uses a microscope. In the portable kit method, a prescribed quantity of oil is passed through a membrane filter of known porous size. The particle counting is done by means of a microscope with the help of a standard contamination specimen. In the microscope method, the particles are sized based on their longest dimension and counted. The limitation of this method is that it does not differentiate the type and size of particles. However, it is quick and gives an on-the-spot assessment; plant personnel can get an idea of where the particles are coming from, i.e., from wear debris or ingress particles, so that suitable corrective measures can be taken.

Automatic particle counting, on the other hand, can quantify the number and size distribution of contaminants. The usual method of automatic particle counting uses light absorption and scattering technique to detect particles suspended in the fluid. A light beam from a laser diode is transmitted through the fluid, passing through the automatic particle counting sensor. Depending upon the particles' size, the fluid sample either absorbs or scatters light, which is detected by a photo sensor; this signal strength is used to determine particle size. This technique is fast and repeatable. Automatic particle counting also helps users evaluate the effectiveness of the filters.

A Note on the ISO Contamination Code When contamination analysis is done with a microscope, the ISO contamination code is based on the number of particles larger than 5 μm and 15 μm present in a 1 mL fluid sample. Once the number and the corresponding size of the particles are determined, this data is converted into scale numbers selected from Table 8–2. The ISO code is specified by two scale numbers separated by a slash, e.g., 18/15. The first number corresponds to the number of particles equal to or greater than 5 μm in size, the second number corresponds to the number of particles equal to or greater than 15 μm in size. For example, in a 1 mL sample, if the number of 5 μm particles is 2,000 and 15 μm particles is 300, then, selecting the appropriate scale numbers from Table 8–2, the fluid cleanliness rating per the ISO code will be 18/15. This practice is in accordance with the old ISO standard 4406-1987.

Per a recently introduced ISO document [3], when the contamination analysis is done by automatic particle counting, the particle sizes are reported by three scale numbers: $\geq$4 μm, $\geq$6 μm and $\geq$14 μm; the last two sizes are equivalent to the old practice of 5 μm and 15 μm particle sizes. For example, in a code of 18/16/13, the first code represents the particles equal to or larger than 4μm, more than 1,300, and up to and including 2,500 particles; the second code represents particles equal to or larger than 6 μm, more than 320, and up to and including 640; the third code represents particles equal to or larger than 14 μm, more than 40, and up to and including 80 in a given 1 mL sample fluid.

Leakage Control

Hydraulic systems are associated with a messy, oil-around image. Neglected leakage control is the cause of that image, and hence leakage control should become a part of a maintenance program. A sudden increase in the topping-up oil consumption or a sudden drop of oil level in the reservoir is an indication that heavy leakage is taking place somewhere. It could be a line burst, seal failure, or hose or pipe fitting failure. Such problems are encountered more frequently in high-pressure systems and so need extra maintenance attention; care must also be taken in the design itself. Leakage can account for 15 to 20 percent of hydraulic oil consumption. Leaked oil can also interfere with the functioning of other lubricants. For example, leaking hydraulic oil, often referred to as tramp oil, will cause splitting of emulsion if it gets mixed with water-based cutting fluids, which makes the fluid unstable and unusable.

Besides technical and economic considerations, leakage oil spilling onto the floor or the machine is anathema to today's cleaner environmental conscious-ness. Improper assembling of fittings, particularly in less accessible places, badly assembled or worn-out seals, improper hose assemblies and mounting plates, and a system subjected to high levels of shock and vibration (causing fittings to loosen) are problems that must be rectified [17]. The quality of fittings, hose assemblies, and seals also has a profound influence on leakage. The extra money spent on quality seals and fittings will always compensate for any money spent on leakage control. Above all, educating personnel about the implications of leakage oil and providing them with appropriate tools will go a long way toward making the system leak free and the workplace clean. Such a step also ensures prevention over cure.

Test Kits

Fluid and equipment manufacturers market test kits for used oil analysis [18–21]. One such portable test kit is shown in fig. 8–9; with this kit, water content, total base number, viscosity, and insoluble content can be measured quickly in situ. In a large production setup, a centralized facility for monitoring the fluid often causes organizational problems and hampers the rhythm of maintenance activities. Test kits provide a decentralized, self-contained facility for mainte-nance and make plant personnel self-driven and self-directed. This will also involve plant personnel in maintenance and will facilitate the implementation of a maintenance schedule without any hiccups. It is cost-effective as well as time-saving.

Oil Changing

Following a sound maintenance schedule will indicate when oil should be changed; procurement of oil and downtime for the machine can therefore be planned, facilitating minimum production disruption. When the oil is changed, the system must be thoroughly flushed to remove the old oxidized oil [22]. As pointed out earlier, partially oxidized oil is an effective catalyst for the oxida-tion of fresh oil. Flushing must be done with the same oil or with flushing oil at about 40°C to 50°C (104°F to 122°F) and should be continued until the filter has no trace of contaminants. Typical flushing time may range from two to eight hours. For large systems of 2,000 liters or more, flushing may run for some

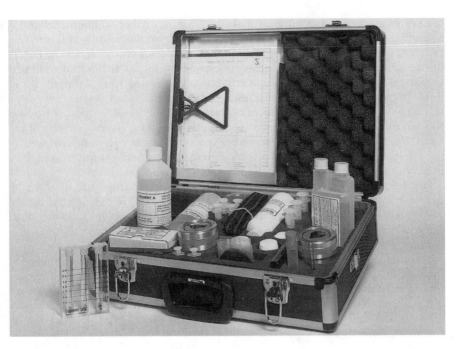

FIGURE 8–9 Portable Oil Test Apparatus (Courtesy: Stanhope-Seta Limited, U.K)

days. It is desirable to flush even a new system to ensure reliable operation of the hydraulics.

Keeping the oil clean not only extends its life, it also ensures trouble-free operation of the hydraulics and long life for the hydraulic components. The cost associated with frequent changes of oil, breakdowns, and replacement of components justifies the extra time and money spent on upkeeping hydraulic oil clean.

MAINTENANCE OF INVERT EMULSION

Invert emulsion is water-based and demands more labor for the upkeep of the fluid; it needs more frequent attention and a well-organized maintenance program. A maintenance program should include a check on water content to ensure the fluid's fire resistance, viscosity, cleanliness, and stability.

Operating Temperature/Water Content/Viscosity

The desirable bulk fluid temperature in the reservoir is 50°C (122°F). However, a value of up to 60°C (140°F) is permitted, taking into account the localized temperature, which can be around 65°C. This temperature limit will help keep the evaporation rate to a minimum. Despite the care taken in maintaining the operating temperature, depletion of water due to evaporation is still a major problem with invert emulsions. Reduction of water quantity reduces the fluids' fire resistance and viscosity—contrary to expected thinking. Incidentally, change in viscosity is also used as a rough guide to knowing the water content of the fluid. To meet fire resistance requirements, the recommended practice is to keep the water content above 35 percent by volume. The precise method of finding the percentage of water content is the traditional acid split test. On the other hand, if a higher viscosity is observed, it is an indication that emulsion split is the likely cause. Hence, significant change in the viscosity of a fluid requires further investigation in order to take suitable remedial measures.

The strength of emulsion should be corrected by adding only distilled or deionized water. When makeup water is added, it should be introduced slowly at the suction side of the pump so that internal churning can help emulsification. Alternatively, water can be added in the form of fine spray when the pump is working.

Another problem with invert emulsion is demulsification. When the evaporated water gets condensed, it does not re-emulsify; instead, it sinks to the bottom of the tank as a separate layer. The presence of free water is checked by using a clean rod covered with water-finding paste. When this rod is dipped into the tank, the color of the rod changes from red to white [23]. If the amount of free water exceeds 4% of the total volume, the fluid should be changed.

Contamination Control

Contamination is more severe with these fluids than with mineral oils because invert emulsion acts as a cleansing agent and the contaminants become suspended instead of settling down. Though some fluids are reported to work with a filter of finer rating, say, 3 μm, the recommended practice is to use filters with a 10 μm nominal rating, as the emulsion is prone to splitting into water and oil. Hence the fluid can easily be contaminated with particle sizes of less than 10 μm. The commonly accepted contamination level is ISO 19/17/14. If the fluid is heavily contaminated with smaller particles, it must be discarded. Periodic checks of contamination are important, as higher contamination levels can adversely affect the life of the pump and other hydraulic components.

Stability of Emulsion

The problem of emulsion instability prevents a wider application of invert emulsion. Emulsion's stability can be disturbed in two ways, apart from improper mixing or poor quality of the fluid or water. First, if the fluid is not in operation for a long period of time—is kept in a stagnated condition—the oil and water can separate. Second, when the fluid is mixed with tramp oil, i.e. other lubricating oils used in the machine, the emulsion can break into water and oil. Tramp oil should be periodically skimmed off from the fluid. If the fluid is prone to tramp oil contamination, it is advisable to use a tramp oil skimmer. Stability of the emulsion can be improved considerably by proper mixing of water and oil. Water should be added slowly to the oil phase under high shear conditions to reduce emulsion particle size and thus attain stability.

Details on the emulsion stability test procedures are given in Chapter 6. A quick in-situ check for emulsion stability is to put drops of an emulsion sample into a container of water. If the emulsion is stable, it will form liquid beads as it plunges below the waterline; on rising to the surface, the beads will not collapse and will remain intact. An unstable fluid, however, will spread throughout the water as a milky fluid [24].

MAINTENANCE OF WATER GLYCOLS

Water Content/Viscosity

Besides routine checks on filter maintenance and leakage controls, monitoring water content and pH levels are part of a glycol maintenance schedule. As with invert emulsion, water content has to be checked periodically, and water content generally should not be allowed to fall below 35 percent. This limit can, however, marginally vary from one supplier to another, and the manufacturer's recommendation should also be taken into account. Unlike invert emulsions, the viscosity of water glycols is inversely proportional to the water content. Any loss in water strength thickens the fluid, which can cause operational problems. The actual percentage of water content can be established by measuring viscosity. Fluid suppliers provide a chart indicating the amount of water added against different values of viscosity. A typical graph showing the viscosity of water glycols as a function of water content is shown in fig. 8–10. Alternatively, a quick in-situ

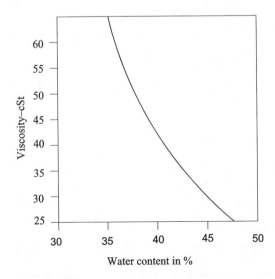

FIGURE 8–10 **Relation between Viscosity and Water Content**

method is to use a pocket-type refractometer (Chapter 4) along with a calibration chart to check the water content. Exact water content can be determined by the distillation method per ASTM D 95 (Chapter 3). As in invert emulsion, only deionized or distilled water should be added to restore the original water ratio.

Monitoring pH

Alkalinity or pH is another fluid parameter that should be monitored periodically. Oxidation causes the alkalinity level to fall, which may impair the corrosion-inhibiting property of the fluid. The normal pH value is around 9.2; the general recommended minimum value is around 8. The initial pH level can be restored by adding the correct quantity of alkaline additives. Fluid suppliers provide the necessary guidelines. The pH value can be measured by means of a pH meter or by the traditional titration method, in which a sample of solution is titrated with a mineral acid to neutralize the alkalinity present in the solution. Alternatively, pH can also be checked by using pH papers.

There are only a few parameters to monitor with water glycols, and so it is quite practicable to implement the checks without any hassles. Because these fluids have a high water content, any lapses in their proper care will impair the functioning of the system as well as cause costly breakdowns.

MAINTENANCE OF PHOSPHATE ESTERS

Phosphate esters are easier to maintain than water-based fluids. But because they are six to eight times more expensive than mineral oils, any lapse in maintenance will lead to high replacement cost. As are mineral oils, phosphate esters are subject to thermal and oxidative degradation. Thermal degradation can produce carbonaceous deposits, and high temperatures will accelerate the oxidation process and give rise to significant changes in viscosity and TAN values. As temperature is one of the main causes of degradation, the bulk fluid temperature should be kept between 50°C and 65°C (122°F to 140°F). The minimum temperature limit of 50°C is kept to check the rise of water content in the fluid.

Monitoring Water Content

Water is the most dangerous contaminant to phosphate esters because they have a strong tendency to hydrolyse. Phosphate esters are subject to autocatalytic hydrolytic degradation, i.e., the acidic products resulting from hydrolytic action act as a catalyst for further hydrolysis. This action causes both corrosion and erosion, and also affects foaming and air release properties. As it can with mineral oil, water can also cause emulsification of these fluids, thus affecting their stability. Therefore it is of prime importance to control the water content of the fluid. The recommended water content limit is below 0.1 percent. Some manufacturers recommend a maxim limit of only 0.05 percent. It is advisable to put a desiccant such as silica gel into the breather as a moisture trap. When the operating environment makes the system susceptible to the easy ingress of water, it may be essential to install a vacuum dehydration unit to periodically separate the water from the fluid. It is advisable to prevent the water ingress problem than to cure it.

Monitoring TAN

Phosphate esters, like mineral oils, are subject to oxidative degradation and in the process become acidic. Temperature and contaminants like water and metal particulate promote oxidation, increasing the value of TAN. Hence monitoring TAN is an important part of maintenance. To control the acidity level, the usual procedure is to treat the fluid with activated clay (Fuller's earth). Fuller's earth is processed from attapulgus clay or attapulgite, which contains alkaline substance such as magnesium oxide and hydroxide. Depending upon the system, the fluid is treated either continuously or at intervals through portable units. Fuller's earth is quite effective in controlling acidity, but it causes the formation of metal soaps. Samples analyzed at intervals over a period of time indicate that solids such as magnesium hydroxide and sodium aluminum silicates present in the clay react with the acidic fluid to form soluble and insoluble calcium and magnesium salts, that could form gels, particularly at low temperatures. This also leads to the formation of black, brittle, solid deposits on hot surfaces such as bearings [25]. Because the acidic fluid promotes autocatalytic hydrolytic dehydration—i.e., the higher the TAN, the higher the rate of hydrolysis—it is imperative to keep the TAN value to as low as 0.1 [26, 27]. As long as the TAN level is maintained within this limit, purification of the fluid with Fuller's earth will adequately maintain the fluid for a long period of time without any adverse effects. When the increase in TAN exceeds the limit of 0.1, the Fuller's earth must be changed. A 1 to 1.5 percent w/v ratio of Fuller's earth/fluid is the recommended dosage, where 5 to 10 percent of the volume is circulated every hour and the preferred mesh size is 50/80 [27, 28].

To overcome the problem of metal deposition, activated alumina is used as an alternative to Fuller's earth. Activated alumina can be natural as well as synthetic. Zeolite, a natural hydrous aluminum silicate of sodium, calcium, potassium, or barium, is a widely used alumina, but synthetically manufactured ones possess consistency in the composition. Activated alumina does alleviate the problem of metal deposition, as it contains no calcium and no magnesium. But the sodium content in alumina can react with highly acidic fluids. While the full effects of activated alumina on the quality of fluid conditioning has yet to be established, alumina is less effective than Fuller's earth in controlling the resistivity, corrosion, and oxidation performance of fresh fluid. The phenomenon of high resistivity (low conductivity) is important particularly when electrohydraulic servo valves are used in the system, as low resistivity causes electrochemical erosion in servo valves [29]. Low resistivity of a fluid is also an indication that the fluid is either acidic or contaminated with water or chlorine. A proprietary material—a composite of alumina and zeolite—controls acidity, minimizing the release of metal salts into the fluid [30].

Table 8–6   Suggested Limit for the Operating Parameters
of Phosphate Esters

Sl. No.	Property	Limit
1	Viscosity @ 40°C	±10%
2	TAN	0.2
3	Water content	1,500 ppm
4	Air release	10 minutes @ 50°C
5	Mineral oil content	0.5%
6	Contamination level	ISO class 18/16/13
7	Calcium/Magnesium	10 ppm of any one
8	Sodium	30 ppm
9	Chlorine content	100 ppm
10	Volume resistivity	40 MΩm @ 20°C

Extract from *Reolube Turbo fluids-A guide to their maintenance and use,* © FMC Corporation. [28]

As Fuller's earth and activated alumina are hygroscopic, they readily absorb moisture from the atmosphere and can accelerate hydrolytic action in the fluid. Hence they should be stored in plastic bags in a dry environment, and should be dried at around 110°C (230°F) for at least 12 hours—but check with the manufacturer that the material can withstand that temperature [28].

To obviate the adverse effects associated with solid absorbent materials, the emerging technique is to treat the fluid with ion-exchange resins [31]. This process can not only condition fluids that are in operation but can also reclaim fluids that are heavily contaminated. In other words, the ion-exchange process can help keep fluids in the system almost indefinitely. Studies have been conducted on whether to use a weak base anionic treatment or a cationic resin treatment. Because synthetic phosphate esters are most widely used, the recommended practice is to use a weak base anionic treatment combined with vacuum dehydration at moderate vacuum and system temperatures so that hydrolysis can be eliminated. Though ion-exchange treatment effectively controls degradation of a fluid, its initial cost is higher than that of other processes. But considering the long life it gives to the fluid, ion-exchange is economical in the long run.

Metal content, electrical resistivity, and chlorine content must also be monitored. Metal in the fluid can come from absorbent media or from acidic degradation products. These metals can lead to the formation of soaps or gels, which can affect foaming and air release properties. Chlorine content must also be controlled, as it causes servo valve erosion [29]. Chlorine usually enters the system through chlorine cleaning solvents or through the cooling water, particularly when it is seawater or hard water. The limits for the most important operating parameters of phosphate esters are given in Table 8–6 [28].

Observation on Existing Practice

When phosphate esters are used in big forging and metal casting applications, particularly in an automated setup, metal dusts are a major source of

Table 8–7   Guide to Fluid-Related Problems

Problem	Checklist	Remedy
Unacceptable temperature rise	Check oil level	Maintain the recommended oil level
	Oil is dumped more frequently through relief valve	Set pressure is high; set the correct pressure. Check at the circuit design–examine whether oil can be by-passed, or consider using variable-delivery pump
	Relief valve may be stuck	Clean the relief valve if required
	Viscosity may be high	Select lower viscosity grade oil
	System may require heat exchanger/heat exchanger undersized	Install heat exchanger/size it correctly
	Blockage in the cooler	Clean the cooler lines
	In piston pumps, pump may not be running full	Pump may need priming, examine whether circuit requires booster pump, suction lines badly engineered, air leaks at suction pipe joints—rectify these defects
Inconsistent feed	Air lock in the system	Bleed the air
	Temperature fluctuation high	Install heat exchanger
	Oil oxidized badly	Check TAN; change oil if TAN is 2 or more
Sluggish movement of slide	Viscosity of the oil may be high	Select lower viscosity grade oil if the viscosity selected is high. Preheat the oil if ambient temperature is low
	Sticky spool valve	Check the fluid condition; clean the fluid and valve
	Incorrect sizing of valves, tubes, pumps, etc.	Size the components correctly
Pump cavitation	Start-up temperature may be low or oil viscosity is too high	Preheat the fluid; select a lower viscosity oil
	Pump starved of fluid–low oil level, suction pressure loss too high, or pump speed too high, clogged strainer	Correct the oil level, reduce pressure losses in suction lines, clean the strainer, and air breather, and reduce the pump speed
Corrosion-related problems	Check for water contamination	Prevent water ingress and dewater the fluid if required
	Check fluid compatibility if copper alloys are involved	Change over to compatible fluid

(continues)

Table 8–7 *(Continued)*

Problem	Checklist	Remedy
Burnt odor or rapid darkening of oil	Seal friction may be high– burnt smell of seals, improper fitting or wrong selection of seals	Correct the seal fitting with the right seal
	Air bubbles in the system	Prevent air ingress; ensure deaeration
Unacceptable level of foaming	Low residence time	Increase the residence time
	Suction pressure too low	Engineer the suction lines properly
	Oil level low in the reservoir or oil badly contaminated	Keep the recommended oil level, clean the oil, or change the oil/switch to an alternative brand
	Air leak at suction/suction pipe joints	Minimize joints, engineer the suction lines properly
	Check the seals in the pump area	If the seals are worn out, replace them
	Check for water contamination	Water content of even 300 ppm sometime can cause foaming; centrifuge the oil if contaminated with water
Too much sludges	High operating temperature, heavily contaminated, oil oxidized badly	Examine the system thoroughly and eliminate the causes; check TAN; change the oil if required

contamination. Dust entering through cylinder wiper seals are a common source of trouble. Wiper seals provided in the rod, it should be remembered, are not a complete barrier against the entry of dust, particularly when the operating environment is very dusty and particles are fine. It is essential to control this contamination by providing bellows in the cylinder rods; the bellow material must be compatible with phosphate esters, and use of a breather bag should also be considered. In such applications, monitoring the contamination level of the fluid becomes an important part of a maintenance schedule. Periodic off-line filtering or centrifuging is a convenient method of purging the contaminants from the fluid, particularly in an automated system, as they will not dislocate regular production activities.

In a continuous production process, found mostly in metal-working industries, production figures often override orderly fluid maintenance. But production targets depend on the reliable operation of hydraulic systems. The common (and inadequate) practice is to change fluids at arbitrary intervals with no regular monitoring of fluid condition. Since these systems are normally large, (with a fluid capacity of at least 6,000 liters, for example) and phosphate esters costly, it is prudent to recognize that extending fluid life by even a small amount can save a considerable amount of money as well as provide a trouble-free production run. Fluid changes should be carried out only after a proper assessment of the fluid condition is done according to the standard test methods. This requires systematic monitoring of fluids.

CONCLUSION

Maintenance is important to the efficient operation of hydraulic systems as well as to the longevity of fluid and equipment. Lubricants are often compared to blood in the human circulatory system, and oil analysis is analogous to blood analysis. Periodic checks of oil condition give clues to health of equipment, which may be warnings to rectify the system, thus averting costly breakdowns. Besides independent labs, oil companies and hydraulic equipment manufacturers also provide oil analysis service [32–34]. This may be a good option for small enterprises. Personnel training is also part of successful maintenance, as human error can be costly. Techniques and tools need constant review and upgrading, not only to make maintenance effective but also to motivate the people who do the work. Yet despite proper maintenance, unforeseen fluid-related problems may crop up, and guidelines to tackle such problems are given in Table 8–7.

Chapter 9

DISPOSAL, REGULATIONS, RECLAMATION, AND REREFINING

The various steps outlined in the previous chapter for the upkeep of hydraulic fluids can appreciably extend their life. Nevertheless, the age of a fluid is still finite; at some point, it becomes used oil. Generation of used oil is a recurring process in industry; managing used oil is very much a part of managing hydraulic systems, and its disposal an important component.

USED OIL: WHAT DOES IT MEAN?

Considering the large number of used-oil sources, it is essential to define the term used oil; otherwise it means different things to different people. There are many definitions for used oils [1–10]. One [1] is any semisolid or liquid used product consisting totally or partially of mineral oil or synthesized hydrocarbons (synthetic oils), oil residues from tanks, oil-water mixtures, and emulsions. These used products result from industrial and nonindustrial sources, where they have been used for lubricating, hydraulic, heat transfer, electrical insulating (dielectric), or other purposes. Their original characteristics have changed during use, thereby rendering them unsuitable for further use. The terms "used oil" and "waste oil" are used synonymously. By this definition, vegetable and animal oils, petroleum-based solvents, waste oils from storage tanks, and spill oils are not considered used oils. Likewise, per the European Economic Community (EEC) directives [3], disposal is defined as "the processing or destruction of waste oils as well as their storage and tipping above or underground."

In the past, used oils were discarded as if they had no value, and their impact on the environment was hardly recognized. But the oil crisis of the 1970s, coupled with growing environmental concerns and tighter government regulations, have changed that scenario. It is imperative that used mineral oils be disposed of in a responsible way, as they can cause considerable damage to the environment. For example, one gallon of used mineral oil can spoil one million gallons of drinking water—a year's supply for 50 people in the U.S. [11]. Also, because it spreads, even a small quantity of mineral oil can significantly damage water streams–one gallon of oil can completely cover one acre (4,047 m^2) of water spread. Indiscriminate dumping of used oils in soil is also a potential threat

Table 9–1   Utilization of Used Oils [14] in Million Liters

Disposal methods	U.S.	Japan	EEC	Canada
Used as fuels	3,070–58%	806–61%	906–37%	76–18%
Reprocessed	113–2%	100–8%	845–34%	98–23%
Reuse	420–8%	Not investigated		Not investigated
Dumping, others	1,700–32%	399–31%	725–29%	250–59%
Total	5,303–100%	1,305–100%	2,476–100%	424–100%

Note: Reuse means used lubricants used as mold releasing agents for concrete, mixing with asphalt for road construction, and pouring on roads for dust suppression

to vegetation. As for the aquatic life, even 0.1 ppm of mineral oil reduces the life span of shrimp by 20 percent [12]. Biodegradability of mineral oil is quite poor (Chapter 7), and hence it should never be dumped in wasteland, drains, streams, or sewers. There are many hazardous contaminants in used hydraulic oils, which have the potential to cause significant environmental damage—polar compounds, chlorine, sulfur, lead, arsenic, barium, and calcium are a few commonly found substances. But when hydraulic oil is mixed with other oils, such as motor oils, solvents, or insulating oils, toxic substances such as polyacrylic aromatics (PCA) and polychlorinated biphenyl (PCB) can be created. Because mineral oils are low volatile and water insoluble, they seep and penetrate into soil, polluting the subsoil waterbed. Mineral oil is the largest nonaqueous liquid hazardous waste in the world.

Among the consumption of lubricants, the ratio of consumption of industrial lubricants and automotive lubricants is roughly 55:45, and this figure can vary by about 10 percent in some countries. Among the consumption of industrial lubricants, hydraulic fluids rank the highest—for example, they account for 39 percent of the consumption of industrial lubricants in the EEC [13]. Though figures for the U.S. and Japan regarding consumption of hydraulic fluids are not available, it is unlikely that they are much different. It is clear that the volume of used hydraulic fluids in any industry is substantial, though it is not as high as that of automotive lubricants. Therefore it is good economic and social sense to integrate disposal activity with other key technical and management activities. The data on the ways used lubricants are dealt with in major lubricant-consuming countries is given in Table 9–1 [14]. This data has awakened the concerned countries to initiate steps to end inappropriate disposal. It is evident that huge volumes of lubricants are disposed of in a way that damages the environment. The combined consumption of lubricants by the U.S., the EEC, and Japan constitutes nearly 42 percent of the world's consumption [13]; these countries have initiated measures to dispose of used oils more responsibly.

GOVERNMENT REGULATIONS

EEC Directives

Government regulation has always played a crucial role in bringing about any transformation in the social order. While the full involvement of society is

Table 9–2   Extract of the European Economic Community (EEC) Directives on Disposal of Waste Oils (Directives 75/439 and 87/101)

- Necessary measures have to be taken to ensure that waste oils are collected and disposed of without causing any avoidable damage to man and the environment. Discharge of waste oils into inland surface water, ground water, territorial seawater and drainage system is prohibited. Deposit or discharge of waste oils or any residues resulting from the processing of waste oils harmful to the soil is prohibited. Any processing of waste oils causing air pollution should not exceed the level prescribed by the existing provisions.
- Priority should be given to the processing of waste oil by regeneration. Where regeneration is constrained by technical, economical and organizational considerations, combustion of waste oils can be carried out under environmentally acceptable conditions and emission level should be within the set limits. Where both are not possible, safe destruction or their controlled storage and tipping should be ensured.
- Any agencies undertaking waste oil collection must be subject to registration and adequate supervision by the competent national authorities and must obtain a permit. Permit shall be granted after examination of the installations.
- The base oils regenerated should not contain toxic and dangerous wastes as defined in the EEC Directive and should not contain polychlorinated biphenyls and polychlorinated tophynyls (PCB/PCT) in concentration beyond a maximum limit, which in no case may exceed 50 ppm.
- The regeneration of waste oils containing PCBs and PCTs is permitted if the regeneration process makes it possible either to destroy the PCBs and PCTs or to reduce them so that the regenerated oils do not contain PCB and PCT beyond the set maximum limit of 50 ppm.
- Waste oils when used as fuel should also not contain PCB/PCT in concentrations beyond 50 ppm.
- During storage and collection, holders and collectors must not mix waste oils with PCBs and PCTs nor with toxic and dangerous waste within the meaning of Directives.

equally essential, it is imperative that government authorities step in and regulate the disposal of used oils to arrest growing environmental degradation. Many regulations exist for the disposal of used lubricants in many countries. The earliest-known government regulations came from Germany, which date back to 1968 and are known as the Waste Oil Laws [15]. EEC regulations appeared in 1975 [2]. The EEC amended this act in 1987 [3], and putting more teeth into it; an extract is given in Table 9–2. EEC regulations are comprehensive and stringent.

Based on the EEC directives, member states have passed their own legislation. Italy, for instance, has its own regulations [4] and has founded a Waste Oil Consortium, which is a legal entity that regulates waste oil disposal. The consortium ensures the collection of waste oils and delivers them to the rerefiner. Where rerefining is not possible, the consortium arranges for other forms of energy-efficient uses or disposes of the oils in compliance with pollution laws. The penalty for the improper disposal of used oils (dumping) in Italy is stiff, carrying a fine as high as ~$10,350 U.S. (20 million Lire) or two years' imprisonment. Italian consumers are also forbidden to eliminate waste oils on their own.

It is inspiring to note the recent judgment by the European Court against Germany for its noncompliance with an EEC directive on the disposal of waste oil [16], this decision clearly reflects the EEC's commitment to the safe disposal of waste oils. On September 9, 1999, the European Court declared that by failing to take the measures necessary to give priority to the processing of waste oils by regeneration, notwithstanding that technical, economic, and organizational constraints were so allowed, the Federal Republic of Germany had failed to fulfill its obligations under Article 3(1) of Council directive 75/439/101/EEC of 16th June 1975 on the disposal of waste oils, as amended by Court Directive 87/101/EEC of 22nd December 1986. Germany was ordered to pay the costs. The EEC is serious.

U.S. Regulations

In the U.S., disposal of used oil is regulated both by federal and state legislation. At the federal level, it is regulated under the authorities of the Resource Conservation and Recovery Act (RCRA) [7, 11, 17]. After the passage of the RCRA in 1978, the Environmental Protection Agency (EPA) initiated discussion over whether or not to list used oil as hazardous waste. In 1980 the Congress, to encourage recycling and expanding the use of recycled oil, passed the Used Oil Recycling Act, which provides financial assistance to states to educate the public about the recycling of used oils. In 1985, the EPA proposed listing all types of used oils as hazardous waste. A year later the EPA withdrew that proposal, presumably thinking it would hamper collection and recycling activities and also lead to improper disposal. In 1988, the issue reappeared when the U.S. District Court for the District of Columbia ordered the EPA to review its decision. In 1992, the EPA reached a final decision: It would not list used oil meant for disposal as hazardous waste. But if used oil were mixed with any listed hazardous waste, it would be subject to hazardous-waste regulations.

The U.S. federal law that is most relevant to the recycling of oil is the Comprehensive Environmental Response, Compensation, and Liability Act (CERCLA, or Superfund Law). Per CERCLA, the agencies that handle used oils—such as used oil generators, haulers, and disposal firms—bear cleanup liability for spilling and contaminating any sites with used oils. Burning of used oils is permitted, provided it meets the EPA industrial furnace limits.

In the U.S., many states have their own waste oil management laws [5–7, 17–23]. Alaska, Arizona, Colorado, Iowa, Nevada, North Carolina, Oklahoma, Oregon, South Carolina, and Texas have followed federal regulations; California, Kansas, Massachusetts, Minnesota, Missouri, New Jersey, Ohio, Rhode Island, and Vermont have stricter regulations and list used oils as hazardous waste [17, 18, 23]. Used-oil generators must comply with the laws and regulations of state/local authorities. These regulations are periodically revised; current guidelines can be obtained from the EPA in Washington, D.C., from the National Technical Information Service (NTIS) in Springfield, Virginia, or from state authorities. The RCRA also provides a hotline (800-424-9346) on waste oil handling.

Government regulation of the disposal of used oils exists in Canada, Australia, South Africa, India, Japan, and South Korea [8, 9, 24, 25].

ASME CODE OF ETHICS

But mere compliance with environmental law is not enough. Industry needs to regulate itself in an ethical and socially responsible way. This point is well enunciated in the American Society of Mechanical Engineers (ASME) Code of Ethics of Engineers:

- Engineers shall hold paramount the safety, health, and welfare of the public in the performance of their professional duties.
- Engineers shall consider environmental impact in the performance of their professional duties.
- Whenever the Engineers' professional judgments are overruled under circumstances where the safety, health, and welfare of the public are endangered, the Engineers shall inform their clients and/or employers of the possible consequences.

Let us observe these ethics in letter and spirit and make state regulations redundant.

DISPOSAL OPTIONS

Before we go into disposal methods, let us remind ourselves of the oft-quoted three Rs: Reduce, Reuse, Recycle. Our primary objective must be to reduce the generation of used oil. Better housekeeping and extending the life of the oil—the contamination, leakage, and temperature controls discussed in the previous chapter—must be followed religiously. The other two Rs, reusing and recycling, will be discussed later.

In an increasingly competitive business environment, disposal costs should not become an added burden to a company. The choice of disposal method depends upon a number of factors, such as the volume of used oil generated and its condition; the criticality of end uses; the company's operational practices; elemental costs such as storage, transportation, investment in equipment cost, and the return on such investment; local regulations; organizational and ecological constraints; etc. All these factors must be carefully weighed against one another.

For small generators of used oil, the easiest and probably the best option is to dispose off to a licensed hauler, which frees them from legal tangles. But users must ensure that used oil is not mixed with any listed hazardous wastes, such as solvents, antifreeze, and gasoline. Environmental considerations aside, it is good business sense to segregate used oil from other hazardous waste, as hazardous waste disposal is lengthy, costly, and subject to complicated regulations that require record keeping, reporting, inspections, transportation, accumulation time, emergency prevention and preparedness, emergency response, etc., to name only a few. Some recyclers pay for the collected used oil as a collection fee, if it becomes hazardous waste, the generator must pay a pick-up fee. So it is imperative that the generator avoid mixing used oil with any hazardous substances. Generators of used oils should verify that the collector/transporter has the valid EPA ID number, or the state or local permit for transporting the used oil off-site.

Burning Used Oil

The high-energy content in used oil encourages its use as a fuel either alone or mixed with other fuel oils. One gallon of used oil processed for fuel contains about 140,000 BTU [26], and it is cheaper than normal fuel oil. Hence, burning used oil as fuel is a popular option. One method is to burn it directly in space heaters at the source by generators, which is often called on-site energy recovery. But this is a questionable environmental practice, as the fuel is generally burned without any pretreatment, and hence the contaminants present in the used oil could produce hazardous waste or emit flue gas containing toxic products. This practice should be discontinued. On–site energy recovery is also regulated [17]. First, the used oil to be burned must be owner-generated. Second, the capacity of the heater must not be more than 0.5 million BTU per hour. Third, the combustion gases from the heater must be vented to the atmosphere.

Off-site energy recovery is a more common form of burning. This is done by oil processors who collect, process, burn, or offer for sale the processed oil as specification fuel. Cement kilns are the usual vessels for burning, as they operate at high temperatures and possibly the burning can be done without any environmental damage. To some extent, boilers and other industrial furnaces are also used to burn used oil. As a fuel it is also used in the road stone industry for drying aggregate. The power-generation and metal-refining industries use it in coking plants, brickworks, stone quarries for drying stone, and smelters handling iron, lead, tin, and aluminum [1, 27].

If the required high temperature is not present, incomplete combustion of used oil can take place; this may cause scaling on burner equipment, which can choke even burner nozzles. Pretreatment of used oil is an essential requirement. The normal pretreatment includes gravity settling, centrifuging, and filtering to remove solid contaminants; chemical treatment, also called demineralization, to remove chemical impurities; vacuum distillation to remove water; and possibly clay treatment to remove unwanted additives. The type of pretreatment to be used depends upon the condition and the volume of the used oil.

Used oil meant for burning must also meet certain quality specifications; the EPA specification per Title 40 Code of Federal Regulation (CFR) Part 279.11 is given in Table 9–3. Oil that exceeds the limit given in the specification is referred to as off-specification used oil. Note that although the prescribed limit for halogens is 4,000 ppm, if the total halogens content is more than 1,000 ppm, it is presumed to be a hazardous waste, and the used-oil handler must rebut that presumption by proving that the used oil does not contain any listed halogenated hazardous waste. Similarly, the PCB content in the used oil to be burned for energy recovery should be zero; otherwise it will be subject to the regulations given in 40 CFR 761.20(e) [17, 21]. Restrictions also exist in Germany; used oil meant for reprocessing must not contain more than 20 mg/kg of PCB and more than 2 g/kg of halogen as a whole [14]. The burning of the oil must comply with federal/state air quality/emission controls and ventilations regulations. Used-oil

Table 9-3   U.S. EPA Used Oil Specifications

Pollutant	Limit
Arsenic	Maximum 5 ppm
Cadmium	Maximum 2 ppm
Chromium	Maximum 10 ppm
Lead	Maximum 100 ppm
Flash point	Minimum 100°F (37.8°C)
Total halogens (fluorine, chlorine, bromine, iodine, and astatine)	Maximum 4,000 ppm

burning can release a host of gases such as sulfur and nitrous oxides, carbon monoxide, and carbon dioxide; volatile organic compounds; and metals such as lead, zinc, arsenic, cadmium, manganese, nickel, and chromium into the atmosphere. Emission controls for waste oil combustion in the EEC are given in Table 9-4 [3, 4]. If required, flue gas emission control devices must be employed in order to comply with emission regulations.

Although burning is an energy recovery process, perceived as a value-added method, there is growing recognition that it must end. New EEC legislation, which is in the draft stage, states that incineration should be the last-resort

Table 9-4   Emission Limits for Combustion of Waste Oils

EEC		Italy
Pollutant	Limit value mg/Nm^3(1)	
Cd	0.5	0.2
Ni	1.0	1.0
either (2)	or (2)	
Cr	Cr	(a)
Cu } 1.5	Cu } 5.0	5.0
V	V	0.2
Pb 5.0	Pb	0.2
Cl (3)	100	–
F (4)	5	–
SO$_2$ (5)	–	–
Dust-Total (5)	–	–

Notes:
1. These limit values, which may not be exceeded when waste oils are burnt, indicate the mass concentration of emission of the aforementioned substance in waste gas, in terms of the volume of waste gas in the standard state (273 K; 1,013 hPa), after deduction of the water vapor moisture content, and of a 3% oxygen content by volume in waste gas. In the EEC this is applicable to plants, whose thermal power capacity is three MW or above but for Italy, it is six MW or above.
2. Member state shall lay down the option in their country.
3. Inorganic gases compounds of chlorine expressed in hydrogen chloride.
4. Inorganic gases compounds of fluorine expressed in hydrogen fluoride.
5. It is not possible to determine limit values for these substances at this stage. Member states can independently set the limit, taking into account Directive 80/779/EEC (O).
(a) Chromium in its hexavalent form must never exceed 1mg/Nm^3

disposal method, and the Recycling Forum in Europe [28] is campaigning to end combustion. In a report made by the Medical Officer of Health in Toronto, Canada, it was concluded that burning is an additional contribution to air pollution and should be discontinued as a matter of public health and environmental policy. Also on health grounds, the Toronto City Council asked the Ministry of the Environment to discontinue the issuance of air approvals for waste oil burners when rerefiners exist and to develop sunset regulation to phase out the use of waste oil burners [8]. In Ontario alone, it is estimated that rerefining could reduce greenhouse gas emissions by 200,000 tons each year [29]. Given the limitations and shortcomings of enforcement authorities, controlling and monitoring environmental pollution caused by used-oil burning may be quite difficult. A survey conducted by the Ministry of Environmental Studies in Ontario revealing that 79 percent of waste oil space burners were out of compliance with their certificates of approval [8] only confirms this point. *Oil never wears out; burning waste oil is burning a valuable resource.*

Other Uses

There are other ways waste oils have been used (or misused): as road oiling to control dust, as a raw material in asphalt production, as a pesticide carrier, weed killer, vehicle undercoating, and as a secondary lubricant [1]. It has also been used in oil refineries as feedstock in small quantities without any pretreatment; this is called slipstreaming, which aids the manufacture of other refined products. Road oiling, weed killer, or other applications of used oil that pollute the soil are now prohibited.

Reclamation

In simple terms, reclamation is a cleaning process. Per the ASTM standard, [10] reclamation is the use of cleaning methods during recycling primarily to remove insoluble contaminants; the methods may include settling, heating, dehydration, filtration, and centrifuging. The oil recovered is generally used for the same application or for topping up, or it is burnt as a fuel depending upon the quality of the recovered oil. The quality of the recovered oil depends on the quality of the used oil. Sometimes these oils are also used as metal cutting fluids.

For industrial lubricants, particularly hydraulic oils, reclamation is a viable way to reuse hydraulic oils. Unlike motor oils, which are more complex in composition and are subject to harsh working conditions, hydraulic oils are simpler and less stressed by temperature and contaminants. They are also better maintained than other industrial lubricants, and hence reclamation of hydraulic oils is a practical and profitable proposition. Reclamation can be done at an in-plant facility, thus reducing disposal costs [30–41]. Many portable reclamation units are available that can be connected to individual machines and can handle 15 gallons (56 liters) to 1,200 gallons (4,540 liters). The portable units usually contain filtering, vacuum dehydration, air stripping, and electrostatic filtration devices; they can effectively remove suspended particles, dissolved and emulsified water, volatile impurities, wear metals, and oxidized products. Alternatively, if more machines are involved, a centralized facility can be created. Reclaimed oil is also tested for its specifications. Based on the test results, the oil is then refortified with suitable additives and can be restored to its original state. A centralized facility can handle an oil capacity of up to 2,000 gallons (7,570 liters).

Major lubricant manufacturers also provide on-site and off-site reclamation and additive refortification services [42–44]. Commercial mobile reclamation units are available to recycle used oils on-site. These mobile units contain different modules, such as a series of filtering units, centrifuges, vacuum dehydrators/vacuum distillation units, and even blending tanks for compounding additives. In fact, depending upon the customer's requirement, different combinations of modules can be incorporated. This is an attractive option, as the user does not have to invest in equipment or maintain it. A mobile unit also ensures users that only their oil is used, so they don't have to worry about contamination from other sources.

General Motors (GM) has accrued many advantages from reclamation: reduction in new oil purchase and disposal costs associated with used oils, as well as reduction in hydraulic-related problems and a consequent decrease in lost production and labor costs that result from reduced maintenance expenses and from reduction in environmental liability [31]. GM reclaims 5 million gallons (18.9 million liters) of oil annually that otherwise would have been burnt as fuel and would have produced 55,000 tons of CO_2 emissions a year [31]. Bowering et al., report that recycling hydraulic oil in their cold mill plant saves them approximately $200,000 a year [41]. The experience of GM and other industrial users [31–41] is proof that reclamation yields great benefits.

Segregation Segregation, an important part of reclamation, is the process of separating the various oils that come from different machines and equipment, and identifying, labeling, and storing them. The quality of reclaimed oil depends on how well the various oils have been segregated. Mixing the oils with other lubricants, coolants, solvents, and other cleaning agents should be strictly prevented. Used hydraulic oils mixed with sulfurized neat cutting oil, for instance, cannot be recycled in-plant or used for the same application [45]. And when low-quality and high-quality oils are mixed, the end product is always low-quality oil. In other words, when LVI oils such as cutting oils are mixed with HVI oils such as hydraulic oils, which can easily happen in industry, the end product will be LVI oils. Mixing various oils can also complicate water separation as well as reclamation itself; sometimes it may even make reclamation impossible. Careless dumping of used oil from a machine's reservoir will negate reclamation. Dedicated storage with a built-in heating facility will help settle water and particulates. The segregated oil then can be suitably reclaimed. Industries using various types of oils can separate used oils into the following categories:

- Hydraulic mineral oils
- Automotive lubricants
- Cutting oils
- Water-based fluids
- Insulating oils

This list could be lengthened or shortened, depending upon the volume of oil generated in each category. Great care must be exercised particularly with insulating oils (transformer oils), as some may contain PCBs. Oils containing PCBs should be separated from all other oils to avoid entering the realm of hazardous waste regulation.

Rerefining

Rerefining is the process by which all contaminants, such as oxidation products, additives, and other soluble and insoluble substances of used oils, are removed to produce lube stocks. In other words, rerefining is the process by which virgin-quality base oil is recovered from used oil.

This is the last disposal option discussed in this chapter, but it is also the best option from many points of view. Per the EPA [26], rerefining used oils takes about one-third the energy needed to refine crude oil to lubricant quality. To produce 2.5 quartz (~2.5 liters) of new high-quality lubricating oil, 42 gallons (159 liters) of crude oil are required, as against only one gallon of used oil. Since "oil never wears out," it is abundantly clear that rerefining is the best option. To a national economy, it provides valuable energy and resource conservation. To a company, it provides lubricants at a lower cost. To humanity, it provides a cleaner environment. Putting all these points together, the idea that energy is recovered through burning (i.e., reused only once) is not rational.

The rerefining process is akin to crude oil refining and so is obviously a capital-intensive process. Many commercial/patented rerefining processes exist; they are given in Table 9-5 [1, 14, 27, 45–56]. Dang gives a good account of commercially available rerefining processes [51]. Most rerefining technology was evolved to meet the market needs of motor oils, as the bulk of used-oil generation comes from the automobile users. Motor oils are more complex and more contaminated than many industrial lubricants, so the rerefining process also has to be complex. The rerefining industry grew slowly for decades, partly because viable quantities of feedstock were unavailable in spite of the huge generation of used automotive lubricants.

A brief account of the various rerefining processes will be given here. Most involve preprocess, which includes settling, filtering, dehydration, fuel stripping, and removal of solid particles. The core process, which is the key, is briefly explained below.

Sulfuric Acid and Clay Treatment This is the oldest process, and because it is not sophisticated, it is the most widely used. In this method, waste oil is treated with 95 to 98 percent concentrated sulfuric acid after dehydration. The acid sludge formed, which may contain metallic particles, spent additives, and carbon and combustion products, is then separated out. The remaining oil is heated and contacted with the clay, neutralized, and filtered. This is a low-yield process that produces inferior oil, environmentally unacceptable waste, and consequent disposal problems. It is being replaced by other processes and is now mostly confined to small operators.

Table 9–5   Major Rerefining Processes and Units

Process name	Preprocess	Core process	Finishing Process	Plants
Sulfuric acid and clay	Settling and dehydration	Sulfuric acid treatment	Clay treatment	Many locations
Thin film evaporation	Settling and dehydration	Thin film low-pressure and high-pressure reduction evaporation	Clay and hydrogenation	Safety-Kleen, Elk Grove Village, IL, U.S.
Evergreen	Chemical treatment	Thin film reduced pressure evaporation	Hydrogenation	Newark, California, U.S. and other locations
Viscolube/IFP (TDA/PDA)	Dehydration	Thermal deasphaltation/ propane extraction	Hydrogenation	Viscolube, Pieve, Fissiraga, Italy, and other locations
Dollbergen	Dehydration	Thin film distillation	Clay treatment	Uetze-Dollbergen, Germany
Interline	Mellon process	Proprietary propane-based solvent extraction	Clay treatment	Salt Lake city, Utah, U.S. and other locations
Orcol	Dehydration	Thin film vacuum distillation	Micro filtration	Orcol, Knowsley, U.K.
Vaxon	Dehydration	Thin film distillation	Potassium hydroxide treatment	Cator, Alcover, Spain
KTI/IFP	Dehydration	Thin film vacuum distillation/ propane extraction	Hydrofinishing	LPC, Marousi, Greece
Iwatani	Dehydration	Spark discharge	Clay (reduced pressure evaporation), deaeration and deodoration	Iwatani Chemical Industries, Shinga-prefecture Japan

Notes:
1. The list given is by no means exhaustive.
2. Some companies are in the process of upgrading their technology and hence the processes given here are not necessarily current.

Thin Film Evaporation This method was developed in the U.S., and relies on evaporation by subjecting the fluid to reduced pressure in a thin film evaporator (TFE). An advantage of this process is that distillation is free from decomposition, polymerization, coking, and fouling, as the maximum temperature maintained in the TFE wall is only for 2–5 seconds [48]. The thin film evaporator distills the liquid at each stage batch by batch and recovers hydrocarbon fractions under vacuum. With multiple evaporators, lube oil of different fractions can be obtained. It is simpler to operate than a fractionation tower and has become a popular technology.

Propane Extraction Propane extraction is basically a solvent extraction process. IFP (Institut de France Petroleum), France and Interline, U.S. use this process. After preprocessing, the propane is mixed with the oil, and the metallic and asphaltic components are separated out. Solvent stripper removes the propane, it is recycled, and the remaining oil and light carbons are distilled.

Thermal De-Asphaltation (TDA) This is an IFP/Viscolube (Italy) technology. After preprocessing, the feedstock is fed into the thermal de-asphaltation column, where, under vacuum distillation, metals, impurities, and asphaltic substances are removed from the oil and collected at the bottom of the column. Gasoline is distilled at the top of the column; at the same time three classes of lubricants of different viscosities are fractionated. This process is done at a temperature of about 360° C (680° F) and at vacuum of 20 Torr. After the TDA process, the residue can be further treated with PDA (propane de-asphalting) technology developed by IFP to increase the yield and quality. This technology is cost-effective considering the quality and quantity of yield and the energy spent.

Evergreen Process Evergreen, a California–based company, developed this process, which combines chemical pretreatment and thin film evaporation. After pretreatment, oil is filtered and distilled. The main process is done in a thin film evaporator. The chemical treatment used in preprocess extends the life of the catalysts used in hydrogenation, reducing operational costs.

Iwatani Process This is a Japanese process [14] that uses the electric spark discharging technique. An electric spark pulsating at an interval of 1/1,200 a second, discharged into a tank containing used oil and graphite or metallic pellets, breaks up the fine particles, which are otherwise difficult to filter, and coagulates them into sizes that can be filtered easily. The oil is subsequently treated with the clay and reduced evaporation method. Low process costs, the ability to adapt to varying grades of waste oil, and pollution-free operation are the advantages of this process.

The post-core process is the finishing process, which is common to all rerefining technologies. Clay treatment and hydrogenation are the most common techniques used. Hydrogenation is the same finishing process normally used in the manufacture of lubricating oils from virgin crude (Chapter 2). The clay process was used for many years. The quality of the base oil recovered from this process is inferior compared with the oil produced from the hydrogenation process. The hazardous environmental waste produced in the clay process

Used oil

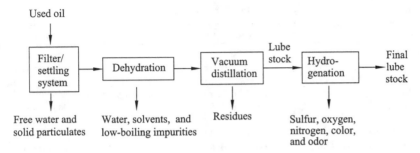

FIGURE 9-1 Typical Steps in Rerefining Process for Mineral Hydraulic Oils

combined with the associated problem of disposal has caused many rerefiners to switch to hydrogenation [27, 47, 48]. Hydrogenation involves the catalytic reaction of hydrogen with oil molecules to remove color, odor, and other auto oxidation elements under high temperature and pressure. Sulfur, oxygen, and nitrogen are also removed, and aromatic substances are saturated in this process. Because hydrogenation is expensive, clay treatment is still in use. Interline claims that its unique clay treatment process, addresses both quality and environmental problems [52, 53]. The various steps in the rerefining process are schematically summarized in fig. 9-1.

The acceptance and growth of rerefined oils was impeded by technological inability to produce virgin quality oils, the high cost of rerefined oils due to poor demand, and an inadequate supply of feedstock to make production economical. But from the variety of technologies discussed above, it is clear that the technology issues have been effectively addressed. Auto majors like Mercedes Benz and Volkswagen have confirmed that the performance of rerefined oils is equal to that of virgin oils. The Technical Association of the European Industry of Lubricants (ATIEL), has redefined the definition of the base stock: "Mineral base stocks may be manufactured from crude oil (virgin base stock) or from used oil (rerefined stock) by refining or rerefining processes." But ATIEL also specified that rerefined stocks must be substantially free from materials introduced through manufacturing contamination or previous use [28]. Groupement Européen de l' Industrie de la Régénération (GEIR) the European Association of Rerefining Industries, has taken another step to assure users of the quality of rerefined oil by formulating a standard for the acceptable quality level for waste oils [28]. This information is given in Table 9-6.

The EEC proposes (in draft) that all public procurement of motor oils should be recycled oils [28]. This is a sweeping measure that would make rerefining economical and probably push the technology further because of the vast market potential. Compared with Europe, apprehension about the quality of rerefined oil persists in the U.S., although not as strongly as a few years ago. As Viscolube has noted, crude oil itself is a recycled product—the transformation of masses of organic matter present in the subsoil into hydrocarbon over thousands of years. Initiatives taken by federal and state authorities and the American Petroleum

Table 9–6   Waste Oil Acceptable Levels Per GEIR

Parameter	Test standard	Unit	Allowable limit
Water	ASTM D 95	% weight	Maximum 10
Specific gravity @ 15°C	ASTM D 1298	gram/cc	Maximum 0.9
Viscosity @ 50°C	ASTM D 445	°E	Minimum 1.8
Sediments	ASTM D 2273	% weight	Maximum 3
PCB/PCT	ASTM D 4059	ppm	Maximum 20
Diluents	ASTM D 322	% volume	Maximum 1
Total chlorine	ASTM D 1317	ppm	Maximum 2,000
Neutralization No. (TAN)	ASTM D 664	mg KOH/g	Maximum 3
Saponification No.	ASTM D 94	mg KOH/g	Maximum 10

GEIR–Groupement Européen de l'Industrie de la Régénération (The European Association of Rerefining Industries)

Institute (API) will, it is hoped, soon relieve public apprehension about the quality of rerefined oils.

Collection of Waste Oils An inadequate supply of used oils to rerefiners on a continuous basis is the main obstacle to the growth of rerefining. How to collect all the waste oils generated? This problem is quantified in Table 9–7 [28, 58, 59], and it is worldwide. In 1993 Watanabe (Japan) said the issue of waste oil treatment is the issue of collecting waste oils [14].

The collection mechanism developed in various places is centered on do-it-yourselfers (DIYs) who change their own oil, mostly vehicle owners. This group generates vast amounts of used oil and also disposes it improperly. The estimated amount of improperly disposed used oil from this source is estimated to be between 193 million to 400 million gallons. In the U.K. it is estimated that DIYs contribute about 1/3 of the total waste oil generated. Many states in the U.S. have embarked upon used-oil collection and oil recycling programs, and Arner has given an excellent review of oil collection programs in various states [55]. To step up oil collection, 17 states have adopted used-oil legislation based on the API model bill known as the Used Oil Collection Act [60]. Creating oil collection funds, grants, subsidies, or loans to local authorities to establish oil collection centers; providing toll-free telephone networks to disseminate information (for

Table 9–7   Used Oil Collected in Various Countries in Tons

Country	Oil consumption per annum	Used oil collected	%
U.S.	8,467,000	4,233,000	50
EEC	4,746,000	1,748,000	37
Japan	2,101,000	882,000	42
Germany	817,000	473,000	58
Italy	600,000	180,000	30
U.K.	800,000	350,000	44
France	890,000	236,000	26
Spain	570,000	130,000	23

example, the location of the nearest oil collection center); and certification of used-oil collection centers are the salient features of this model bill.

In line with this model bill, the oil collection programs of most states are funded by a small tax levied on the sale of new oil. In an estimate by API, there are 26,021 public and private collection centers across the U.S. apart from the 12,200 drop-off centers operated by API and independent dealers. Leading lube manufacturers also provide used-oil collection, as part of comprehensive fluid management services to their customers. Evergreen collects used oil from more than 6,000 sites in California by operating 60 trucks; the sources of the used oil are quite varied—DIYs, auto workshops, fleet maintenance depots, and governmental and industrial locations [56]. Another leading rerefiner, Safety-Kleen, collects approximately 250 million gallons (946 million liters) of used oil annually and produces 80 million gallons (303 million liters) of lube base oil. Many provinces in Canada have also launched similar oil recycling and used-oil collection programs [8, 29].

The collection method adopted in South Africa is worth mentioning here [25]. The Rose (Recovery of Oil Saves the Environment) Foundation, a nonprofit organization, uses three sizes of specially designed containers to collect used oils. Depending upon the volume of used oil produced, a generator can select one of these tanks and install it on her or his premises. So far 7,600 tanks have been installed in South Africa, which can hold 11.5 million liters of used oil. The cost of the tank must be borne by the generator, but it can be adjusted against the earnings from the used oil collected. Sealed couplings are provided in the tank to facilitate direct pumping of oil from the tank to the truck without any spillage. The used oil collected is transported to the approved recyclers. The system is quite simple to operate, and in its three years of operation, the Rose Foundation has significantly increased the volume of collection. Today the rerefined base oil meets 10 to 15 percent of South Africa's base oil requirement.

The consumption of lubricants in growing economies such as India's may not be high at present, but as industrialization speeds up, that consumption is likely to grow at an exponential rate. Before the used-oil problem takes on monstrous dimensions in developing nations, it is time to take concrete recycling measures. This is not just an environmental issue but a serious economic issue as well. Fluctuating crude oil prices always cause economic ripples in countries such as India, and recycling used oil can significantly reduce reliance on oil imports, which in turn can significantly relieve pressure on the economy.

In Japan, waste oil is collected through various sources: gas stations, automobile garages, industrial sites, oil storage terminals, transport depots, government establishments, and other sources. Of the 718 million liters collected, 46.5 percent of the oil collected came from industrial sources, possibly because this is an organized sector, while motor oil collection (gas stations and auto workshops) constituted 37 percent. Even in Japan, Watanabe reports, on-site and off-site burning (energy recovery) accounts for 70 percent of the waste oil generated; hardly 3 percent is recycled [61]. Lack of government support is one cause for the low percentage of recycled oil in Japan. In contrast, waste oil collection and rerefining activities are gaining ground in South Korea, as the government is keen to promote it. In Korea, a small tax is included in the cost of virgin oil to support these activities.

SPECIFIC TO HYDRAULIC OILS

First, the possibility of reclaiming used hydraulic oils on-site or off-site should be thoroughly examined. Next, disposing it with a collector who is a licensee of a rerefiner should be considered. As a last resort, used oil can be given to any licensed hauler, but never in a manner that is unacceptable to government regulations or the environment.

Hydraulic oils are normally much less contaminated than other lubricants, and a rerefining unit with a process specifically chosen for hydraulic oils and possibly used for other industrial lubricants like turbine oils, spindle oils, and gear oils, could spring up as exclusive units. Moreover, these units may not need to be as complex as current ones and may require fewer steps, which could reduce capital investment, operational costs, and plant size itself. Even with a modest volume of used oils, the process could become economical. Collection of the required volume of used hydraulic oils may also not be much of a problem, as these oils are mostly generated in industry. In such a situation more rerefining units could be installed, facilitating easy access to used-oil generators. This option, of course, presupposes that oil contamination is controlled both at the generators' and collectors' ends. In fact, creating an exclusive facility to recycle hydraulic oils could catalyze the idea of rerefining in industry, which could bring a viable volume of used oils on a continuous basis to the rerefiner. If the required technology and the commitment of all concerned people–used-oil generators, contractors, and rerefining personnel–could work in harmony, the benefits of rerefined hydraulic oils could be harnessed to the advantage of society at large.

DISPOSAL OF HWCF AND INVERT EMULSION

The high water content in used fluids does not mean they can be discharged into drains, sewers, streams, or wasteland because invert emulsion or HWCF contain mineral oil with additives, which can significantly damage the environment. It is essential to split fluid into oil and water. Splitting of emulsion can be done by either the physical or chemical method. The usual physical methods are sedimentation, centrifuging, coalescence, and ultrafiltration.

Ultrafiltration is a well-established separation process [62, 63] that is akin to reverse osmosis. In this process, the fluid is passed through a membrane with a pore size of six to seven microns at a pressure of about 170 kPa (25 psig). The emulsified oil, suspended particles, bacteria, etc., are retained in the membrane, and the treated water passes out through the membrane. In order to make this process more effective as well as to relieve the load on the membrane, the fluid is often pretreated. Magnetic separators, paper filtration, centrifuging, tramp oil skimmers, etc., are some of the devices used for pretreatment. Pretreatment devices must be selected according to the nature of the contaminants. Because only a small volume of oil is recovered, it can be easily incinerated or disposed of as explained earlier. The water separated from this process is free of environmentally hazardous substances. Treated water from any physical process that is discharged into sewage systems, however, must meet pollution regulations.

The usual chemical process is acid splitting. Here, hydrochloric acid is added to the fluid to split the emulsion. Oil recovered from chemical process can be recycled or used as fuel.

Another alternative disposal method for HWCF is: it can be mixed with water and disposed of as waste effluent, if local regulation permits. HWCF can also be incinerated, though it has a high water content; it is mixed with more flammable fluid and incinerated, which is however, a slow process.

Equipment to separate the emulsion into water and oil is available commercially [64–67].

DISPOSAL OF WATER GLYCOLS

The disposal problem is little easier with glycols, as they do not leave oil film on surfaces because of their good water solubility. These fluids may be subjected to ultrafilteration or distillation; the glycol can be recovered and water disposed of. Glycols should never be dumped into streams, as it needs a great deal of biological oxygen to degrade, which harms aquatic life.

DISPOSAL OF PHOSPHATE ESTERS

Incineration is the usual way to dispose of phosphate esters. Incineration requires high-temperature combustion facilities, which are uncommon. Combustion in a cement kiln is one possible source of incineration. Otherwise the fluid can be reclaimed by distillation or filtration and used for less demanding applications.

DISPOSAL OF BIODEGRADABLE HYDRAULIC FLUIDS

The term "biodegradable" does not mean that these fluids can be dumped anywhere. The best one can say is that these fluids are less damaging to the environment than mineral oils are. While it's possible that a small amount of used biodegradable fluid getting into a sewage drain may not have serious environmental impact, indiscriminate dumping is an entirely different matter. So far, studies and tests have centered on new fluids; there is no real data yet on the behavior of used biodegradable oils in the ground or in aquatic environments, particularly in the long term. Biodegradable fluids should never be dumped into sewage drains or wasteland. Disposal of these fluids should also follow the commonly accepted practices.

Unlike mineral oils, biodegradable fluids cannot be rerefined. Reclamation services may, however, be available from the supplier. Biodegradable fluids must always be segregated from other petroleum-based or synthetic lubricants. Government regulations about disposal must be observed. Germany, for example, regulates the disposal of biodegradable fluids, which are classified as "critical garbage." These fluids can be incinerated or disposed of through a contractor. In conclusion, the precautions, rules, and care observed for the disposal of mineral oils must be observed for biodegradable fluids as well.

CONCLUSION

Used oil is a valuable resource. No longer used oil could be reused or discarded in ways that are wasteful and hazardous to the environment. Current momentum in environmental protection and tougher government regulations will further tighten the improper disposal of used oils. Consequently, the cost of used oil disposal is going to be dearer. Also, availability and the cost of petroleum crude are not going to remain same forever. There is no guarantee that oil crisis witnessed in the 1970s will not repeat itself in the coming years. Industry must recognize these factors and gear up to manage the used oil in a responsible and rational way—a step that is proactive rather than crisis driven. In this regard, reclamation and rerefining of used oil are good options, as they address both social and economic issues effectively. Reclamation and rerefining of used oils should become a widespread industrial practice. Let us take conscious effort and strive together toward a cleaner environment and resource conservation.

EPILOGUE

I took my course in tribology at Leeds in the mid-1970s. The subject of tribology then, subsequent to the Jost committee report, was vibrant and much talked about both in academia and industry. The focus of attention was to take tribology out of the libraries and bring it to design and machine floors to promote better tribology practice in industry. Training and educational programs sprung up. A great deal of information was distilled into a form that could be used by designers with ease and confidence. Industry began to see the economic significance of tribology. Yet despite the many educational and training programs that were created, the prevailing attitude of industry did not change enough to significantly improve tribological practice. In other words, the tribology revolution that started in the late 1960s lost its steam. Probably, the chip revolution to some extent overshadowed the tribology revolution. There is also a feeling among mechanical engineers that delving into the subject of lubricants and lubrication, as opposed to subjects like mechatronics, controls, manufacturing, or nanotechnology, may not be as rewarding in terms of esteem or money.

In fact, lubrication is rarely discussed or addressed in industry. This situation recalls the famous, "The Oil Shed Fallacy" narrated by the celebrated tribologist Mayo Hersey in the 1930s. He asked: "Why call upon doctors of science to study the shearing stresses in steel shafting, leaving the shearing stresses in the oil film to the doctors of janitorial service?" The situation since then, has changed albeit not much. Professional lubrication engineers are yet to be entrusted to upgrade the art of selecting lubricants from a private game of chance between oil salesman and purchasing agents to a rational process.

Tribological knowledge must be applied to the problems encountered in engineering industries, and a professional tribologist is the person to do this. The current practice, however, only reflects Mayo Hersey's saying: "Mike, the oil shed in charge has the only real dope in the organization, and that he learned it from the labels on the oil barrels." For years, tribology societies, forums, and journals have addressed this issue. Still, the gap between tribologists and industry continues to exist. This book is a small effort toward closing that gap. If readers find the book has helped to demystify and explain hydraulic fluids, then it has served its purpose. If it convinces industry that tribology helps an organization run efficiently and cost effectively, then I have been successful.

Biodegradable fluids and the disposal and rerefining of used oils are the critical issues facing all those concerned with hydraulic fluids. Users must take initiative to propel the development of biodegradable fluids, no longer leaving it in the hands of researchers and lube manufacturers. This book should help users to stretch thinking on this subject. As for disposal and rerefining, much work has been already done; what is needed is the proverbial "extra mile." Again, user industries can pursue this issue aggressively so that energy saving and a cleaner environment can become a reality worldwide.

There are different ways information can be presented to practicing engineers to facilitate better tribological practice. This book has used one method: basic theory combined with practical information and data. This has been done in the hope that readers can travel all the way through the problem, rather than just putting one foot in the road. Readers are welcome to share their experience so that disseminating information to practicing engineers can become more userfriendly.

Tribology is fundamental to many branches of engineering and will continue to be so; it is essential to new technologies such as micromechanisms, mechatronics, nanotechnology, and microelectronics. The drive to make things whether that thing is a gem clip, a computer chip or a spaceship, is part of human nature. Tribology is ingrained into these activities; lubricants will always be a part of human lives. Hence the mission of tribology will never end. I am happy to be part of that mission and will be happier if this book serves it further.

M. Radhakrishnan

Appendix 1

DATA SHEETS ON ANTIWEAR MINERAL HYDRAULIC OILS–TYPE HM

ACKNOWLEDGMENT

The data sheets given are extracts from the *BFPA Report P12-1996*. Permission granted by the British Fluid Power Association is gratefully acknowledged.

Note: Data given are per British Standard unless otherwise specified and values given are only typical. Some companies are in the process of upgrading and rationalizing their products. For the current data or additional data, readers may contact the concerned fluid suppliers.

Table A1–1

184

Sl. No.	Manufacturer	ALEXANDER DUCKHAM			
	Brand	Zircon 15	Zircon 32	Zircon 46	Zircon 68
1	Viscosity grade	15	32	46	68
2	ISO oil type	HM	HM	HM	HM
3	Viscosity cSt @ 0°C	90	250	470	775
	40°C	15	32	46	68
	100°C	3.32	5.4	6.9	9.0
4	Viscosity index	84	108	105	105
5	Temperature °C at which oil has a viscosity between 800 and 1,000 cSt	−26	−12	15	2
6	Relative density @ 15/15°C	0.861	0.876	0.879	0.882
7	Pour point °C	−45	−30	−30	−30
8	Flash point °C open cup (COC)	162	216	225	240
9	Water separation–ASTM D 1401, minutes	*	*	10	*
10	Foaming tendency/stability, mL Seq 1	–	5/nil	10/nil	10/nil
	Seq 2	–	20/nil	30/nil	30/nil
	Seq 3	–	5/nil	10/nil	10/nil
11	Air release–minutes to 0.2% air @ 50°C	–	3.0	–	–
12	Neutralization number, mg KOH/g	0.5	0.5	0.5	0.5
13	Copper corrosion–3 hours @ 100°C	1B	1B	1B	1B
14	Rust prevention–Procedure A distilled water	–	–	–	–
	Procedure B salt water	Pass	Pass	Pass	Pass
15	Acidity change after oxidation @ 1,000 hours– ASTM D 943, mg KOH/g	*	*	0.16	*
16	Seal compatibility index–ISO 6072	–	16	13	10
17	Vane pump wear test– total ring, and vane weight loss, mg	*	*	18	*

*Additives same as for Zircon 46

Table A1-2

Manufacturer				BP OIL			
Sl. No.	Brand	Bartran 22	Bartran 32	Bartran 46	Bartran 68	Bartran 100	Bartran 150
1	Viscosity grade	22	32	46	68	100	150
2	ISO oil type	HM	HM	HM	HM	HM	HM
3	Viscosity cSt @ 0°C	–	–	–	–	–	–
	40°C	22	32	46	68	100	150
	100°C	4.2	5.4	6.8	8.7	11.4	14.6
4	Viscosity index	90	100	100	100	100	95
5	Temperature °C at which oil has a viscosity between 800 and 1,000 cSt	–19	–12	–5	2	8	15
6	Relative density @ 15/15°C	0.875	0.876	0.879	0.882	0.879	0.888
7	Pour point °C	–30	–30	–30	–30	–30	–24
8	Flash point °C open cup (COC)	192	216	225	240	246	267
9	Waters separation–ASTM D 1401, minutes	5	5	10	10	10	–
10	Foaming tendency/stability, mL Seq 1	10/nil	5/nil	5/nil	5/nil	5/nil	5/nil
	Seq 2	50/nil	15/nil	10/nil	10/nil	10/nil	10/nil
	Seq 3	60/nil	5/nil	5/nil	5/nil	5/nil	5/nil
11	Air release–minutes to 0.2% air @ 50°C	–	1.8	–	–	–	–
12	Neutralization number, mg KOH/g	0.22	0.22	0.22	0.22	0.22	0.22
13	Copper corrosion-3 hours @ 100°C	1B	1B	1B	1B	1B	1B
14	Rust prevention–Procedure A distilled water	–	–	–	–	–	–
	Procedure B salt water	Pass	Pass	Pass	Pass	Pass	Pass
15	Acidity change after oxidation @ 1,000 hours– ASTM D 943, mg KOH/g	*	0.2	*	*	*	*
16	Seal compatibility index–ISO 6072	15	13	10	7	6	4
17	Vane pump wear test– total ring, and vane weight loss, mg	*	Approx. 30	*	*	*	*

*Additives same as for Bartran 32

185

Table A1-3

Sl. No.		BP OIL					
	Manufacturer						
	Brand	HLP-HM 22	HLP-HM 32	HLP-HM 46	HLP-HM 68	HLP-HM 100	HLP-D 46
1	Viscosity grade	22	32	46	68	100	46
2	ISO oil type	HM	HM	HM	HM	HM	HM
3	Viscosity cSt @ 0°C	150	250	470	775	1,480	–
	40°C	21	32	46	68	105	46
	100°C	4.2	5.4	6.9	9.0	12.0	6.8
4	Viscosity index	104	108	105	105	101	100
5	Temperature °C at which oil has a viscosity between 800 and 1,000 cSt	–20	–12	–5	2	8	–5
6	Relative density @ 15/15°C	0.875	0.876	0.879	0.882	0.886	0.878
7	Pour point °C	–30	–30	–30	–30	–24	–24
8	Flash point °C open cup (COC)	192	216	225	240	246	225
9	Water separation–ASTM D 1401, minutes	*	*	10	*	*	*
10	Foaming tendency/stability, mL Seq 1	5/nil	5/nil	10/nil	10/nil	20/nil	50/nil
	Seq 2	20/nil	20/nil	30/nil	30/nil	30/nil	30/nil
	Seq 3	6/nil	5/nil	10/nil	10/nil	10/nil	<5/nil
11	Air release–minutes to 0.2% air @ 50°C	–	3.0	–	–	–	–
12	Neutralization number, mg KOH/g	0.5	0.5	0.5	0.5	0.5	0.9
13	Copper corrosion-3 hours @ 100°C	1B	1B	1B	1B	1B	1A
14	Rust prevention–Procedure A distilled water Procedure B salt water	Pass	Pass	Pass	Pass	Pass	–
15	Acidity change after oxidation @ 1,000 hours–ASTM D 943, mg KOH/g	*	*	0.16	*	*	–
16	Seal compatibility index–ISO 6072	16	15	13	10	7	–
17	Vane pump wear test–total ring, and vane weight loss, mg	*	*	18	*	*	–

* Additives as for HLP–HM 46

Table A1–4

Sl. No.	Manufacturer Brand	CASTROL Hyspin AWS 22	Hyspin AWS 32	Hyspin AWS 46	Hyspin AWS 68	Hyspin AWS 100	Hyspin AWS 150
1	Viscosity grade	22	32	46	68	100	150
2	ISO oil type	HM	HM	HM	HM	HM	HM
3	Viscosity cSt @ 0°C	190	350	600	1,000	1,800	3,200
	40°C	22	32	46	68	100	150
	100°C	4.25	5.29	6.65	8.53	11.1	14.5
4	Viscosity index	95	95	95	95	95	95
5	Temperature °C at which oil has a viscosity between 800 and 1,000 cSt	−19	−12	−6	0	4	10
6	Relative density @ 15/15°C	0.875	0.875	0.88	0.88	0.89	0.89
7	Pour point °C	−30	−30	−21	−21	−21	−21
8	Flash point °C open cup (COC)	183	228	249	252	285	288
9	Water separation—ASTM D 1401, minutes	10	10	10	10	15	15
10	Foaming tendency/stability, mL Seq 1	10/nil	trace/nil	trace/nil	trace/nil	trace/nil	trace/nil
	Seq 2	20/nil	20/nil	10/nil	10/nil	trace/nil	trace/nil
	Seq 3	10/nil	trace/nil	trace/nil	trace/nil	trace/nil	trace/nil
11	Air release—minutes to 0.2% air @ 50°C	3	4	7	10	10	20
12	Neutralization number, mg KOH/g	1	1	1	1	1	1
13	Copper corrosion–3 hours @ 100 °C	1	1	1	1	1	1
14	Rust prevention–Procedure A distilled water	No rust	No rust	No rust	No rust	No rust	No rust
	Procedure B salt water	No rust	No rust	No rust	No rust	No rust	No rust
15	Acidity change after oxidation @ 1,000 hours–ASTM D 943, mg KOH/g	<1.5	<1.5	<1.5	<1.5	<1.5	<1.5
16	Seal compatibility index–ISO 6072	11	9	6	5	4	4
17	Vane pump wear test–total ring, and vane weight loss, mg	*	<100	<100	<100	*	*

* Additives as Hyspin AWH 32, 46, and 68

187

Table A1-5

Sl. No.	Manufacturer / Brand	Centraulic CL 22	Centraulic CL 32	Centraulic CL 46	Centraulic CL 68	Centraulic CL 100
				CENTURY OILS		
1	Viscosity grade	22	32	46	68	100
2	ISO oil type	HM	HM	HM	HM	HM
3	Viscosity cSt @ 0°C	184	313.6	524.4	968.5	1,823.5
	40°C	21.5	31.3	44.5	66.5	100
	100°C	4.2	5.4	6.8	8.7	11.1
4	Viscosity index, minimum	90	90	90	90	90
5	Temperature °C at which oil has a viscosity between 800 and 1,000 cSt	−18	−11	−5	2	9
6	Relative density @ 15/15°C	0.87	0.873	0.873	0.88	0.883
7	Pour point °C	−27	−24	−27	−27	−24
8	Flash point °C closed cup	200	205	213	214	210
9	Water Separation–ASTM D 1401, minutes	–	–	–	–	–
10	Foaming tendency/stability, mL Seq 1	–	–	–	–	–
	Seq 2	–	–	–	–	–
	Seq 3	–	–	–	–	–
11	Air release–minutes to 0.2% air @ 50°C	0.4	0.4	0.4	0.4	0.4
12	Neutralization number, mg KOH/g					
13	Copper corrosion–3 hours @ 100°C	1A	1A	1A	1A	1A
14	Rust prevention–Procedure A distilled water	Pass	Pass	Pass	Pass	Pass
	Procedure B salt water	Pass	Pass	Pass	Pass	Pass
15	Acidity change after oxidation @ 1,000 hours– ASTM D 943, mg KOH/g					
16	Seal compatibility index–ISO 6072	–	–	–	–	–
17	Vane pump wear test– total ring, and vane weight loss, mg	<100*	<50*	<50	<50*	<50*

*<50 on ISO 46; others read across

Table A1-6

Sl. No.	Manufacturer / Brand	CENTURY OILS				
		Centraulic AF 22	Centraulic AF 32	Centraulic AF 46	Centraulic AF 68	Centraulic AF 100
1	Viscosity grade	22	32	46	68	100
2	ISO oil type	HM	HM	HM	HM	HM
3	Viscosity cSt @ 0°C	174.8	319.3	595.6	944.3	1,654.9
	40°C	21.5	31	46.5	66.5	100
	100°C	4.3	5.3	6.8	6.8	11.6
4	Viscosity index	100	100	100	100	100
5	Temperature °C at which oil has a viscosity between 800 and 1,000 cSt	−19	−11	−4	1	8
6	Relative density @ 15/15°C	0.865	0.873	0.873	0.86	0.883
7	Pour point °C	−25	−25	−25	−25	−25
8	Flash point °C closed cup	200	205	210	215	230
9	Water separation–ASTM D 1401, minutes	3	5	7	8	10
10	Foaming tendency/stability, mL Seq 1	10/0	0/10	0/0	0/0	0/0
	Seq 2	20/0	10/0	40/0	5/0	0/0
	Seq 3	10/0	0/10	0/0	0/0	0/0
11	Air release–minutes to 0.2% air @ 50°C	3	5	7	10	–
12	Neutralization number, mg KOH/g	–	–	–	–	–
13	Copper corrosion–3 hours @ 100°C	1A	1A	1A	1A	1A
14	Rust prevention–Procedure A distilled water	Pass	Pass	Pass	Pass	Pass
	Procedure B salt water	Pass	Pass	Pass	Pass	Pass
15	Acidity change after oxidation @ 1,000 hours– ASTM D 943, mg KOH/g	–	–	–	–	–
16	Seal compatibility index–ISO 6072	–	–	–	–	–
17	Vane pump wear test– total ring, and vane weight loss, mg	<100*	<50*	<50	<50*	<50*

*<50 on ISO 46; others read across

Table A1–7

Sl. No.	Manufacturer Brand	ESSO					
		Nuto H 22	Nuto H 32	Nuto H 46	Nuto H 68	Nuto H 100	Nuto H 150
1	Viscosity grade	22	32	46	68	100	150
2	ISO oil type	HM	HM	HM	HM	HM	HM
3	Viscosity cSt @ 0°C	85	317	525	917	1,585	2,780
	40°C	21.51	30.8	44	64.8	95.1	151
	100°C	4.21	5.27	6.61	8.45	10.74	14.62
4	Viscosity index	100	101	101	100	96	95
5	Temperature °C at which oil has a viscosity between 800 and 1,000 cSt	−18	−12	−5	0	6	12
6	Relative density @ 15/15°C	0.867	0.871	0.875	0.879	0.882	0.886
7	Pour point °C	−36	−33	−30	−27	−24	−24
8	Flash point °C open cup (COC)	222	225	226	234	263	265
9	Water separation–ASTM D 1401, minutes	5	5	5	5	5	5
10	Foaming tendency/stability, mL Seq 1	0/0	0/0	0/0	0/0	0/0	0/0
	Seq 2	0/0	0/0	0/0	0/0	0/0	0/0
	Seq 3	0/0	0/0	0/0	0/0	0/0	0/0
11	Air release–minutes to 0.2% air @ 50°C	<1	1	3	5	–	–
12	Neutralization number, mg KOH/g	0.66	0.66	0.66	0.66	0.66	0.66
13	Copper corrosion–3 hours @ 100°C	1	1	1	1	1	1
14	Rust prevention–Procedure A distilled water	Pass	Pass	Pass	Pass	Pass	Pass
	Procedure B salt water	Pass	Pass	Pass	Pass	Pass	Pass
15	Acidity change after oxidation @ 1,000 hours– ASTM D 943, mg KOH/g	–	–	–	–	–	–
16	Seal compatibility index–ISO 6072	5	3	3	3	2	–
17	Vane pump wear test– total ring, and vane weight loss, mg	–	<50	<50	<50	–	–

190

Table A1-8

	Manufacturer	FINA						
Sl. No.	Brand	Hydran LZ 22	Hydran LZ 32	Hydran LZ 46	Hydran LZ 68	Hydran LZ 100	Hydran LZ 150	Hydran LZ 220
1	Viscosity grade	22	32	46	68	100	150	220
2	ISO oil type	HM	HM	HM	HM	HM	HM	HM
3	Viscosity cSt @ 0°C	190	320	700	970	1,700	2,980	5,000
	40°C	22	32	46	68	100	150	220
	100°C	4.3	5.2	6.9	8.8	11.3	14.7	19.5
4	Viscosity index	95	100	100	100	100	100	99
5	Temperature °C at which oil has a viscosity between 800 and 1,000 cSt	−18	−11	−5	1	7	14	20
6	Relative density @ 15/15°C	0.865	0.872	0.87	0.881	0.885	0.892	0.895
7	Pour point °C	−35	−35	−35	−35	−25	−25	−20
8	Flash point °C closed cup	190	190	200	200	210	210	220
9	Water separation–ASTM D 1401, minutes	–	–	–	–	–	–	–
10	Foaming tendency/stability, mL Seq 1	5/0	5/0	5/0	5/0	5/0	5/0	5/0
	Seq 2	–	–	–	–	–	–	–
	Seq 3	–	–	–	–	–	–	–
11	Air release–minutes to 0.2% air @ 50°C	5	5	7	10	10	15	15
12	Neutralization number, mg KOH/g	1.5	1.5	1.5	1.5	1.5	1.5	1.5
13	Copper corrosion–3 hours @ 100°C	1	1	1	1	1	1	1
14	Rust prevention–Procedure A distilled water	–	–	–	–	–	–	–
	Procedure B salt water	Pass	Pass	Pass	Pass	Pass	Pass	Pass
15	Acidity change after oxidation @ 1,000 hours– ASTM D 943, mg KOH/g	–	–	–	–	–	–	–
16	Seal compatibility index–ISO 6072	8	7	7	7	7	7	5
17	Vane pump wear test– total ring, and vane weight loss, mg	–	–	–	–	–	–	–

Table A1–9

Sl. No.	Manufacturer	HOUGHTON VAUGHAN					
	Brand	Hydro-Drive HP 22	Hydro-Drive HP 32	Hydro-Drive HP 46	Hydro-Drive HP 68	Hydro-Drive HP 100	Hydro-Drive HP 150
1	Viscosity grade	22	32	46	68	100	150
2	ISO oil type	HM	HM	HM	HM	HM	HM
3	Viscosity cSt @ 0°C	150	300	560	1,000	1,800	3,000
	40°C	22	3.2	46	6.8	100	150
	100°C	4.7	5.6	6.8	8.7	11.2	15.5
4	Viscosity index	100	100	100	100	100	100
5	Temperature °C at which oil has a viscosity between 800 and 1,000 cSt	−11*	−9	−9	0	4	12
6	Relative density @ 15/15°C	0.866	0.871	0.876	0.881	0.886	0.89
7	Pour point °C	−11	−11	−10	−10	−10	−10
8	Flash point °C open cup (COC)	185	210	215	230	240	250
9	Water separation–ASTM D 1401, minutes	10	10	15	20	–	–
10	Foaming tendency/stability, mL Seq 1	40/0	40/0	40/0	40/0	40/0	40/0
	Seq 2	40/0	40/0	40/0	40/0	40/0	40/0
	Seq 3	50/0	50/0	40/0	50/0	50/0	50/0
11	Air release–minutes to 0.2% air @ 50°C	–	–	–	–	–	–
12	Neutralisation number, mg KOH/g	1.4	1.4	1.4	1.4	1.4	1.4
13	Copper corrosion–3 hours @ 100°C	1	1	1	1	1	1
14	Rust prevention–Procedure A distilled water	Pass	Pass	Pass	Pass	Pass	Pass
	Procedure B salt water	Pass	Pass	Pass	Pass	Pass	Pass
15	Acidity change after oxidation @ 1,000 hours– ASTM D 943, mg KOH/g	<2	<2	<2	<2	<2	<2
16	Seal compatibility index–ISO 6072	14	11	10	9	9	9
17	Vane pump wear test– total ring, and vane weight loss, mg	<100	<100	<100	<100	<100	<100

*Pour point

Table A1–10

Sl. No.	Manufacturer		MOBIL			
	Brand	DTE 22	DTE 24	DTE 25	DTE 26	DTE 27
1	Viscosity grade	22	32	46	68	100
2	ISO oil type	HM	HM	HM	HM	HM
3	Viscosity cSt @ 0°C	175	310	510	900	1,600
	40°C	21	30.4	43.5	64.2	93.4
	100°C	4.2	5.23	6.55	8.4	10.8
4	Viscosity index	102	102	101	100	99
5	Temperature °C at which oil has a viscosity between 800 and 1,000 cSt	−19	−12	−6	0	6
6	Relative density @ 15/15°C	0.862	0.873	0.878	0.881	0.887
7	Pour point °C	−30	−24	−21	−21	−15
8	Flash point °C open cup (COC)	186	216	220	230	240
9	Water separation–ASTM D 1401, minutes	6	10	15	15	20
10	Foaming tendency/stability, mL Seq 1	20/0	20/0	20/0	20/0	20/0
	Seq 2	20/0	20/0	20/0	20/0	20/0
	Seq 3	20/0	20/0	20/0	20/0	20/0
11	Air release–minutes to 0.2% air @ 50°C	3	4	6	8	6
12	Neutralization number, mg KOH/g	1.2	1.2	1.2	1.2	1.2
13	Copper corrosion-3hrs @ 100°C	1b	1b	1b	1b	1b
14	Rust prevention–Procedure A distilled water	Pass	Pass	Pass	Pass	Pass
	Procedure B salt water	Pass	Pass	Pass	Pass	Pass
15	Acidity change after oxidation @ 1,000 hours– ASTM D 943, mg KOH/g	–	0.1	–	0.2	–
16	Seal compatibility index–ISO 6072	12	11	8	6	6
17	Vane pump wear test– total ring, and vane weight loss, mg	*	<50	*	*	*

*Additives package as for DTE 24

193

Table A1-11

Sl. No.	Manufacturer	MORRIS LUBRICANTS					
	Brand	Liquimatic No. 3	Liquimatic No. 4	Liquimatic No. 5	Liquimatic No. 6	Liquimatic No. 7	Liquimatic No. 8
1	Viscosity grade	22	32	46	68	100	150
2	ISO oil type	HM	HM	HM	HM	HM	HM
3	Viscosity cSt @ 0°C	160	320	590	950	1,800	3,350
	40°C	20	31	46.5	65	100	140
	100°C	4.2	5.15	6.75	8.4	11.2	13.9
4	Viscosity index	100	98	98	98	98	95
5	Temperature °C at which oil has a viscosity between 800 and 1,000 cSt	−15	−10	−5	0	5	10
6	Relative density @ 15/15°C	0.869	0.873	0.878	0.878	0.882	0.887
7	Pour point °C	−38	−30	−30	−30	−25	−17
8	Flash point °C open cup (COC)	220	228	240	248	260	272
9	Water separation–ASTM D 1401, minutes	3	3	3	3	4	8
10	Foaming tendency stability, mL Seq 1	5/0	5/0	5/0	5/0	5/0	5/0
	Seq 2	20/0	20/0	20/0	20/0	20/0	20/0
	Seq 3	5/0	5/0	5/0	5/0	5/0	5/0
11	Air release–minutes to 0.2% air @ 50°C	3	4	4	8	10	10
12	Neutralization number, mg KOH/g	<0.2	0.2	<0.2	<0.2	<0.2	<0.5
13	Copper corrosion–3 hours @ 100°C	1	1	1	1	1	1
14	Rust prevention–Procedure A distilled water	Pass	Pass	Pass	Pass	Pass	Pass
	Procedure B salt water	Pass	Pass	Pass	Pass	Pass	Pass
15	Acidity change after oxidation @ 1,000 hours–ASTM D 943, mg KOH/g	<1	<1	<1	<1	1	1
16	Seal compatibility index–ISO 6072	12	5	5	3	3	2
17	Vane pump wear test–total ring, and vane weight loss, mg	<50	<50	<50	< 50	<50	<50

Table A1-12

Sl. No.	Manufacturer	SHELL							
	Brand	Tellus 22	Tellus 32	Tellus 37	Tellus 46	Tellus 68	Tellus 100	Tellus R 22	Tellus R 68
1	Viscosity grade	22	32	32/46	46	68	100	22	68
2	ISO oil type	HM	HM	HM	HM	HM	HM	HM	HM
3	Viscosity cSt @ 0°C	180	338	440	580	1,040	1,790	180	1,011
	40°C	22	32	37	46	68	100	22	68
	100°C	4.3	5.3	5.7	6.7	8.6	11.1	4.3	8.7
4	Viscosity index	100	99	99	98	97	96	100	100
5	Temperature °C at which oil has a viscosity between 800 and 1,000 cSt	−15	−12	−10	−7	2	5	−15	2
6	Relative density @ 15/15°C	0.866	0.875	0.875	0.879	0.886	0.891	0.869	0.877
7	Pour point °C	−30	−30	−30	−30	−24	−24	−30	−30
8	Flash point °C open cup (COC)	216	222	224	230	235	249	222	234
9	Water Separation–ASTM D 1401, minutes	15	20	20	20	20	25 @ 82°C	15	20
10	Foaming tendency/stability, mL Seq 1	10/0	10/0	20/0	20/0	30/0	30/0	10/0	30/0
	Seq 2	15/0	15/0	15/0	15/0	20/0	20/0	15/0	20/0
	Seq 3	20/0	10/0	20/0	20/0	30/0	40/0	30/0	30/0
11	Air release–minutes to 0.2% air @ 50°C	4	5	6	7	9	12	4	9
12	Neutralization number, mg KOH/g	0.7	0.7	0.7	0.7	0.7	0.7	0.1	0.1
13	Copper corrosion–3 hours @ 100°C	Class 1	Class 1	Class 1	Class 1	Class 1	Class 1	Class 1	Class 1
14	Rust prevention–Procedure A distilled water	Pass	Pass	Pass	Pass	Pass	Pass	Pass	Pass
	Procedure B salt water	Pass	Pass	Pass	Pass	Pass	Pass	Pass	Pass
15	Acidity change after oxidation @ 1,000 hours– ASTM D 943, mg KOH/g	*	*	<1	*	*	*	*	*
16	Seal compatibility index–ISO 6072	11	9	9	8	6	4	11	9
17	Vane pump wear test–total ring, and vane weight loss, mg	*	*	<15	*	*	*	#	#

*Tested for Tellus 37–comparable performance expected for other VG oils
#Tested for Tellus R37 (<50)–comparable performance expected for other VG oils

195

Table A1–13

		SILKOLINE LUBRICANTS					
	Manufacturer						
Sl. No.	Brand	Derwent 22	Derwent 32	Derwent 46	Derwent 68	Derwent 100	Derwent 150
1	Viscosity grade	22	32	46	68	100	150
2	ISO oil type	HM	HM	HM	HM	HM	HM
3	Viscosity cSt @ 0°C	240	335	620	1,150	1,900	3,320
	40°C	22.05	32	46.1	68.1	100.2	154
	100°C	4.20	5.41	6.87	8.74	11.08	14.8
4	Viscosity index	101	103	104	100	96	96
5	Temperature °C at which oil has a viscosity between 800 and 1,000 cSt	−17	−11	−5	1	8	14
6	Relative density @ 15/15°C	0.867	0.869	0.877	0.88	0.885	0.888
7	Pour point °C	−35	−29	−27	−25	−22	−20
8	Flash point °C closed cup	190	193	200	210	215	220
9	Water separation–ASTM D 1401, minutes	*	*	15	*	*	*
10	Foaming tendency/stability, mL Seq 1	–	–	0/0	–	–	–
	Seq 2	–	–	0/0	–	–	–
	Seq 3	–	–	0/0	–	–	–
11	Air release–minutes to 0.2% air @ 50°C	–	–	–	–	–	–
12	Neutralization number, mg KOH/g	*	*	0.6	*	*	*
13	Copper corrosion–3 hours @ 100°C	–	–	–	–	–	–
14	Rust prevention–Procedure A distilled water	–	–	Pass	–	–	–
	Procedure B salt water	–	–	Pass	–	–	–
15	Acidity change after oxidation @ 1,000 hours– ASTM D 943, mg KOH/g	–	–	–	–	–	–
16	Seal compatibility index–ISO 6072	–	–	–	–	–	–
17	Vane pump wear test– total ring, and vane weight loss, mg	*	*	9	*	*	*

*Not tested but uses same performance package as Derwent 46

Table A1-14

Sl. No.	Manufacturer / Brand	Rando HD 22	Rando HD 32	Rando HD 46	Rando HD 68	Rando HD 100	Rando HD 150
					TEXACO		
1	Viscosity grade	22	32	46	68	100	150
2	ISO oil type	HM	HM	HM	HM	HM	HM
3	Viscosity cSt @ 0°C	178	310	559	1,033	1,862	3,429
	40°C	22	32	46	68	100	150
	100°C	4.4	5.6	6.9	8.7	11	14.1
4	Viscosity index	100	100	100	95	90	90
5	Temperature °C at which oil has a viscosity between 800 and 1,000 cSt	−18.2 −20.4	−11.5 −13.6	−4.1 −6.5	1.8 0.5	9.1 6.6	14.8 12.9
6	Relative density @ 15/15°C	0.873	0.876	0.882	0.887	0.896	0.9
7	Pour point °C	−42	−30	−30	−30	−18	−15
8	Flash point °C open cup (COC)	196	196	204	218	218	240
9	Water separation—ASTM D 1401, minutes	–	–	–	–	–	–
10	Foaming tendency/stability, mL Seq 1	0/0	0/0	0/0	0/0	0/0	0/0
	Seq 2	25/0	50/0	50/0	50/0	50/0	50/0
	Seq 3	0/0	0/0	0/0	0/0	0/0	0/0
11	Air release—minutes to 0.2% air @ 25°C/50°C	5/–	5/–	–/2	–/10	–/10	–/15
12	Neutralization number, mg KOH/g	–	–	–	–	–	–
13	Copper corrosion—3 hours @ 100°C	–	–	–	–	–	–
14	Rust prevention—Procedure A distilled water	Pass	Pass	Pass	Pass	Pass	Pass
	Procedure B salt water	Pass	Pass	Pass	Pass	Pass	Pass
15	Acidity change after oxidation @ 1,000 hours—ASTM D 943, mg KOH/g	–	–	–	–	–	–
16	Seal compatibility index—ISO 6072	120	120	120	120	120	120
17	Vane pump wear test—total ring, and vane weight loss, mg	30	30	30	30	30	30

Table A1–15

Sl. No.		Manufacturer			TOTAL OIL GREAT BRITAIN			
	Brand	Azolla ZS 22	Azolla ZS 32	Azolla ZS 46	Azolla ZS 68	Azolla ZS 100	Azolla ZS 150
1	Viscosity grade	22	32	46	68	100	150
2	ISO oil type	HM	HM	HM	HM	HM	HM
3	Viscosity cSt @ 0°C	175	299	532	1,025	1,655	2,900
	40°C	22	32	46	68	100	150
	100°C	4.4	5.5	6.7	8.9	11.2	15
4	Viscosity index	105	102	100	100	100	100
5	Temperature °C at which oil has a viscosity between 800 and 1,000 cSt	−18	−11	−5	2	8	14
6	Relative density @ 15/15°C	0.871	0.870	0.877	0.884	0.886	0.89
7	Pour point °C	−25	−24	−21	−18	−18	−15
8	Flash point °C open cup (COC)	170	210	230	240	250	250
9	Water separation–ASTM D 1401, minutes	≤25	≤25	≤25	≤25	≤25	≤25
10	Foaming tendency/stability, mL Seq 1	–	10/0	10/0	0/0	0/0	–
	Seq 2	–	10/0	10/0	20/0	20/0	–
	Seq 3	–	10/0	10/0	10/0	10/0	–
11	Air release–minutes to 0.2% air @ 50°C	<5	<5	<5	<5	<5	<5
12	Neutralization number, mg KOH/g	≤1.5	≤1.5	≤1.5	≤1.5	≤1.5	≤1.5
13	Copper corrosion–3 hours @ 100°C	≤1B	≤1B	≤1B	≤1B	≤1B	≤1B
14	Rust prevention–Procedure A distilled water	Pass	Pass	Pass	Pass	Pass	Pass
	Procedure B salt water	Pass	Pass	Pass	Pass	Pass	Pass
15	Acidity change after oxidation @ 1,000 hours– ASTM D 943, mg KOH/g	0.7	0.7	0.7	0.7	0.7	0.7
16	Seal compatibility index, ISO 6072	–	–	–	–	–	–
17	Vane pump wear test– total ring, and vane weight loss, mg	<50	<50	<50	<50	<50	<50

Table A1–16

SI. No.	Manufacturer / Brand	FUCHS Renolin B 5	Renolin B 10	Renolin B 15	Renolin B 20
1	Viscosity grade	22	32	46	68
2	ISO oil type	HM	HM	HM	HM
3	Viscosity cSt @ 0°C	180	350	600	950
	40°C	22	33	47	67
	100°C	4.4	5.6	7.1	8.6
4	Viscosity index	100	100	100	100
5	Temperature °C at which oil has a viscosity between 800 and 1,000 cSt	−22	−12	−8	0
6	Relative density @ 15/15°C	0.865	0.88	0.88	0.882
7	Pour point °C	−30	−27	−27	−24
8	Flash point °C open cup (COC)	168	190	220	225
9	Water separation–ASTM D 1401, minutes	< 15	< 15	< 15	< 15
10	Foaming tendency/stability, mL Seq 1	0/0	0/0	0/0	0/0
	Seq 2	0/0	0/0	0/0	0/0
	Seq 3	0/0	0/0	0/0	0/0
11	Air release–minutes to 0.2% air @ 50°C	3	5	7	8
12	Neutralization number, mg KOH/g	0.3	0.3	0.3	0.3
13	Copper corrosion–3 hours @ 100°C	1-100 A 24	1-100 A 24	1-100 A 24	1-100 A 24
14	Rust prevention–Procedure A distilled water	0-A	0-A	0-A	0-A
	Procedure B salt water	–	–	–	–
15	Acidity change after oxidation @ 1,000 hours–ASTM D 943, mg KOH/g	<2.0	<2.0	<2.0	<2.0
16	Seal compatibility index–ISO 6072	Report	Report	Report	Report
17	Vane pump wear test–total ring, and vane weight loss, mg	<50	<150	<150	<150

Note: This data sheet was provided by M/S Fuchs Petrolube AG and is not an extract from *BFPA Report P12-1996*

Appendix 2

DATA SHEETS ON FIRE RESISTANT HYDRAULIC FLUIDS

ACKNOWLEDGEMENT

The data sheets given are extracts from the *BFPA Report P13-1996*.
Permission granted by the British Fluid Power Association is gratefully acknowledged.
Data on Fuchs and Reolube were provided by Fuchs Petrolube AG, Germany and FMC Corporation (U.K.) respectively.

Note: Data given are per British Standard unless otherwise specified and values given are only typical. Some companies are in the process of upgrading and rationalizing their products. For the current data or additional data, readers may contact the concerned fluid suppliers.

Table A2–1 Invert Emulsion–HFB

	Manufacturer	CASTROL		CENTURY		ESSO
Sl. No.	Brand	Anvol WO 68	Anvol WO 100	Aquacent light	Aquacent heavy	Hydraulic fluid FR-B-100
	Color	White emulsion	White emulsion	Opaque cream emulsion	Opaque yellow emulsion	Opaque white emulsion
1	Viscosity grade	68	100	68	100	100
2	Viscosity kinematic cSt @ 20°C	158	241	168	240	–
	@ 40°C	64	94	68	96	92
	@ 60°C	34	48	33.4	46.5	–
3	Viscosity index	>175	>175	–	–	–
4	Relative density 15/15°C	0.94	0.94	0.954	0.958	0.930
5	Pour point °C	–40	–34	–34	–34	–
6	Water content % volume	40	40	41	41	42.5
7	Emulsion stability–IP290					
	Free oil, mL	1	2	1.5	2	–
	Free water, mL	nil	nil	0.5	trace	–
	Change in water content @ 425 mL level	nil	nil	1%	1%	–
	Change in water content @ 125 mL level	+0.3	+0.5	1%	1%	–
8	Foaming tendency/stability, mL Seq. 1	5/nil	5/nil	10/nil	10/nil	–
9	Anticorrosive qualities–CETOP R 48H					
	Steel	3.4	3.4	Pass	Pass	–
	Copper	11	11	Pass	Pass	–
	Brass	4.6	4.6	Pass	Pass	–
	Zinc	0.9	0.9	Pass	Pass	–
	Aluminum	1.4	1.4	Pass	Pass	–
10	Fire resistance–CEC 6th report					
	Spray ignition para 3.2.2	<10 seconds not ignited	<10 seconds not ignited	Pass	Pass	Pass
	Wick test para 3.3.2	5.1	4.7	Pass	Pass	Pass
11	Seal compatibility index–ISO 6072	1.9	1.8	As for mineral oils		–
12	Effect on paint–report	Epoxy paint preferred		Contact supplier		As for mineral oils
13	Specific heat kJ/kg °C @ 25°C and 1,013 m bar	2.9	2.9	–		–
14	Coefficient of thermal expansion (volumetric)/°C	–	–	0.00059	0.00059	–
15	Fluid maintenance procedure	Water content must not fall below 35% volume. Use tap water		Adjust water content when it falls below 35%		Add clean water when water content falls below 35%

Table A2–2 Invert Emulsion–HFB

Sl. No.		Manufacturer	MOBIL		SHELL		FUCHS	
	Brand		Pyrogard D	Pyrogard C	Irus Fluid BLT 68	Irus Fluid BLT 100	Aquacent light	Aquacent heavy
	Color		Milky white	Milky white	White opaque	Yellow	Cream	Yellow
1	Viscosity grade		100	68	68	100	68	100
2	Viscosity kinematic cSt @ 20°C		300	–	167	239	131 @ 25°C	200 @ 25°C
	@ 40°C		117	68	70	97	68	91
	@ 60°C		–	–	37	46	34	43
3	Viscosity index		190	–	190	187	–	–
4	Relative density 15/15°C		0.924	0.93	0.934	0.933	~1 (IP 160)	~1 (IP 160)
5	Pour point °C		–34	–30	–30	–27	–	–
6	Water content % volume		43	44	42	42	42	42
7	Emulsion stability–IP290						Stable	Stable
	Free oil, mL		5	–	2.5	2		
	Free water, mL		nil	–	nil	nil		
	Change in water content @ 425 mL level		3	–	0.3	nil		
	Change in water content @ 125 mL level		3	–	0.3	–0.3		
8	Foaming tendency/stability, mL Seq. 1		25	25/nil	30/nil	30/nil	7 days/70°C	7 days/70°C
9	Anticorrosive qualities–CETOP R 48H							
	Steel		Pass	Pass	Pass	Pass	Pass	Pass
	Copper		Pass	Pass	Pass	Pass	Pass	Pass
	Brass		Pass	Pass	Pass	Pass	Pass	Pass
	Zinc		Pass	Pass	Pass	Pass	–	–
	Aluminum		Pass	Pass	Pass	Pass	Pass	Pass
10	Fire resistance–CEC 6th report						British coal 570/81	British coal 570/81
	Spray ignition para 3.2.2		3 seconds	–	Pass	Pass	570/81	570/81
	Wick test para 3.3.2		11 seconds	–	Pass	Pass	Pass	Pass
11	Seal compatibility index–ISO 6072		As for mineral oils		As for mineral oils		As for mineral oils	
12	Effect on paint–report		Epoxy or vinyl		Mostly as for mineral oils		Report	
13	Specific heat kJ/kg °C @ 25°C and 1,013 m bar		2.65		–	2.8	3.0 @ 20°C	3.0 @ 20°C
14	Coefficient of thermal expansion (volumetric)/°C		0.00054	0.00054	0.0005	0.0005	–	–
15	Fluid maintenance procedure		Add clean water when water content is below 35% volume		Maintain water content 35% to 45%; add distilled or deionized water only.		Used oil analysis	

203

Table A2–3 Water Glycol–HFC

Sl. NO.		BP OIL		INTERNATIONAL SPECIALITY CHEMICALS		WM CANNING	
Manufacturer							
Brand		BP Enersyn SF-C 14	BP Enersyn SF-C 15	Breox Hydrolube NF46/2180	NF46/2181	Erifon 818	Trustol 88-46
Color		Yellow	Yellow	Amber	Amber	Green	Pink
1	Viscosity grade	46	46	46	46	15	46
2	Viscosity kinematic cSt @ 20°C	99	112	103	99	34	90
	40°C	46	46	46	46	15	46
	50°C	34	34	–	–	–	–
	60°C	–	–	24	25	8	26
3	Viscosity index	–	–	–	–	140	170
4	Relative density 15/15°C	1.075	1.08	1.083	1.076	1.06	1.08
5	Pour point °C	–47	–40	–40	–47	–50	–47
6	Water content % volume	43	36	36	43	36	36
7	Foaming tendency/stability, mL Seq. 1	20/nil	20/nil	20/nil	20/nil	80/nil	100/nil
8	Air release–minutes to 0.2% air @ 50°C	10	10	20	20	5	8
9	pH	9.7	9.7	9.6	9.6	9.4	9.4
10	Anticorrosive qualities–CETOP R 48H						
	Steel	Pass	Pass	Pass	Pass	–	–
	Copper	Pass	Pass	Pass	Pass	–	–
	Brass	Pass	Pass	Pass	Pass	–	–
	Zinc	Pass	Pass	Pass	Pass	–	–
	Aluminum	Pass	Pass	Pass	Pass	–	–
11	Fire resistance–CEC 6th report						
	Spray ignition para 3.2.2	Pass	Pass	Pass	Pass	–	Pass similar Factory Mutual Research Test
	Wick test para 3.3.2	–	–	Pass	Pass	–	–
12	Seal compatibility index–ISO 6072	Use Buna or neoprene		Not compatible–board, paper, and cork		Cork and leather not compatible	
13	Effect on paint–report	Removes most paints		Use epoxy and phenolic resin-based paints		Use epoxy resin-based paint	
14	Specific heat kJ/kg °C @ 25°C and 1,013 m bar	3.3	3.2	–	–	3.35	3.35
15	Coefficient of thermal expansion (volumetric)/°C	–	–	–	–	0.0002	0.0002
16	Fluid maintenance procedure	Maintain water content @ 43%. Use distilled or deionized water		Maintain water content @ 36%. Use distilled or deionized water		Use distilled or deionized water for makeup	

Table A2–4 Water Glycol–HFC

		CASTROL	CENTURY	CENTURY	ESSO	GULF	MOBIL
Sl. NO.	Manufacturer / Brand	Anvol WG 46	FRX Medium	Glycent 346	Hyd. Fluid FR-C 46	FR Fluid WG	Nyvac No. 20
	Color	Hazy red	Colorless, hazy	Clear hazy red	Straw	Water white	Orange-red
1	Viscosity grade	46	46	46	46	46	46
2	Viscosity kinematic cSt @ 20°C	114	90	100	103	100	99
	40°C	46	46	44.5	46	43	46
	50°C	–	–	–	34	–	–
	60°C	24	26.7	23.5	20	22	24
3	Viscosity index	>200	180 minimum	–	165	190	170
4	Relative density 15/15°C	1.07	1.075	1.088	1.08	1.054	1.08 maximum
5	Pour point °C	–51	–26	–35	–40	–26	–30
6	Water content % volume	36	45	35	36	45	43
7	Foaming tendency/stability, mL Seq. 1	10/nil	15/nil	20/nil	20/nil	20/nil	20/nil
8	Air release–minutes to 0.2% air @ 50°C	7	20	20	–	10	–
9	pH	9.2	9.6	9.6	9.6	8.5	9.6
10	Anticorrosive qualities–CETOP R 48H						
	Steel	0.5	Pass	Pass	–1	0.6	Pass
	Copper	1.2	Pass	Pass	–1	0.33	Pass
	Brass	1.2	Pass	Pass	–1	0.9	Pass
	Zinc	1.3	Pass	Pass	+1	9.3	Pass
	Aluminium	1.1	Pass	Pass	–1	0.1	Pass
11	Fire resistance–CEC 6th report						
	Spray ignition para 3.2.2	Not ignited	Pass	Pass	Pass	Pass	Pass
	Wick test para 3.3.2	Not ignited	Pass	Pass	Pass	Pass	Pass
12	Seal compatibility index–ISO 6072	0.71	–	–	Use Nitrile, Viton, PTFE	–	–
13	Effect on paint–report	Vinyl or epoxy resin-based paint	Contact supplier		Use epoxy resin-based paint	Use epoxy resin-based paint	Use epoxy resin –based paint
14	Specific heat kJ/kg °C @ 25°C and 1,013 m bar	3.3	–	–	3.2	3.1	3.3
15	Coefficient of thermal expansion (volumetric)/°C	–	–	–	0.0002	0.007	0.0002
16	Fluid maintenance procedure	Use deionized or distilled water to maintain water content	Use distilled or deionized water for makeup		Check water % and alkalinity	Makeup with deionized water	Use distilled or deionized water for makeup

Table A2-5 Water Glycol–HFC

Sl. NO.		Manufacturer	FUCHS	SHELL	SMALLMAN	QUAKER CHEMICAL	
	Brand		Hydrotherm 48 M	WG Fluid C	Crown Furnasafe No. 2	Quintolubric 720	Quintolubric 730
	Color		–	Clear yellow	Clear red	Colorless to light yellow	
1	Viscosity grade		46	–	–	46	46
2	Viscosity kinematic cSt @ 20°C		105	94	77	98	93
	40°C		47.4	40	39	47	46
	50°C		34.0	–	27	26	26
	60°C		26.0	20	–	–	–
3	Viscosity index		200	140	>150	230	245
4	Relative density 15/15°C		1.085	1.08	–	1.071	1.076
5	Pour point °C		–42	–48	–50	–48	<–38
6	Water content % volume		45	43.5	40	45	45
7	Foaming tendency/stability, mL Seq. 1		90/01	Nil/nil	90/nil	40/nil	40/nil
8	Air release–minute to 0.2% air @ 50°C		20	20	–	25	25
9	pH		10	8.2	9.6	9.4	9.3
10	Anticorrosive qualities–CETOP R 48H						
	Steel		<+1	–	Compatible with steel, copper, brass, and anodized aluminium	Pass	Pass
	Copper		+1	–		Pass	Pass
	Brass		<+1	–		Pass	Pass
	Zinc		>–1	–		Pass	Pass
	Aluminium		<+1	–		Pass	Pass
11	Fire resistance–CEC 6th report		Approved as per 7th Luxembourg report				
	Spray ignition para 3.2.2			Pass	–	Pass	Pass
	Wick test para 3.3.2		Report	Pass	–	Pass	Pass
12	Seal compatibility index–ISO 6072		Report	As for mineral oils	Use butyl, viton, natural rubber, nitrile, neoprene	Use Buna N, viton, PTFE	
13	Effect on paint–report		Report	Use epoxy resin-based paint	Use epoxy, phenolic, vitreous enamel paint	Contact supplier	
14	Specific heat kJ/kg °C @ 25°C and 1,013 m bar		3.27 @ 30°C	2.8	3.15		3.2 @ 20°C
15	Coefficient of thermal expansion (volumetric)/°C		0.00051 @ 20°C	–	0.0007	–	
16	Fluid maintenance procedure		Used oil analysis	Use distilled or deionized water only. Maintain 35% to 45% water	Use distilled or deionized water for makeup	Contact supplier	

206

Table A2-6 Water Glycol–HFC

		HOUGHTON VAUGHAN				
Manufacturer						
Sl. NO.	Brand	Houghton-safe 73	Houghton-safe 105 CF	Houghton-safe 200X	Houghton-safe 620	Houghton-safe 273 CF
	Color	Green	Red	Green	Red	Red
1	Viscosity grade	10	22	46	46	46
2	Viscosity kinematic cSt @ 20°C	25	49	90	90	90
	40°C	10	22	39	39	38.5
	50°C	–	–	–	–	–
	60°C	5	12	23	23	23
3	Viscosity index	140	150	140	140	140
4	Relative density 15/15°C	1.045	1.05	1.05	1,081	1,072
5	Pour point °C	–30	–32	–45	–51	–39
6	Water content % volume	52	52	45	36	45
7	Foaming tendency/stability, mL Seq. 1	–	–	190/nil	230/nil	10/nil
8	Air release–minutes to 0.2% air @ 50°C	–	–	–	–	–
9	pH	9.3	9.3	9.3	9.5	9.5
10	Anticorrosive qualities–CETOP R 48H					
	Steel	Pass	Pass	Pass	Pass	Pass
	Copper	Pass	Pass	Pass	Pass	Pass
	Brass	Pass	Pass	Pass	Pass	Pass
	Zinc	Pass	Pass	Pass	Pass	Pass
	Aluminium	Pass	Pass	Pass	Pass	Pass
11	Fire resistance–CEC 6th report					
	Spray ignition para 3.2.2	Pass	Pass	Pass	Pass	Pass
	Wick test para 3.3.2	Pass	Pass	Pass	Pass	Pass
12	Seal compatibility index–ISO 6072			Compatibility chart available		
13	Effect on paint–report			Compatibility chart available		
14	Specific heat kJ/kg °C @ 25°C and 1,013 m bar	1	1	3.4	3.4	3.4
15	Coefficient of thermal expansion (volumetric)/°C	0.0007	0.0007	0.0007	0.0007	0.0007
16	Fluid maintenance procedure			Maintenance data sheet available		

Table A2–7 Phosphate Esters HFD–R

Sl. NO.		BP OIL	CASTROL		CENTURY		ESSO
	Manufacturer	BP OIL	CASTROL		CENTURY		ESSO
	Brand	BP Energol SF-D46	Anvol PE 46 HR	Phosphate Ester NTP	Fosfocent 430	Fosfocent 600	Hyd. fluid FR-D-46
	Color	Red	Straw	Clear pale green	Clear blue	Pale yellow	Straw
1	Viscosity grade	46	46	–	22	–	46
2	Viscosity kinematic cSt @ 20°C	–	160	240	47	135	300
	40°C	43	46	60.4	22	41.5	42
	60°C	–	18	25.3	12.2	17.5	24
3	Viscosity index	45	< 0	48	179	45	–
4	Temperature in °C for viscosity of 850 cSt	0	4	7	–30	6	5
5	Maximum operating bulk fluid temperature °C	150	Refer to supplier	150	150	150	150
6	Relative density @ 15/15°C	1.127	1.15	1.128	1.123	1.123	1.135
7	Pour point °C	–21	–21	–25	–34	–18	–23
8	Foaming tendency/stability, mL	20/nil	5/nil	25/nil	50/nil	20/nil	100/nil
9	Air release–minutes to 0.2% air @ 50°C	–	5	7	5	7	2
10	Neutralization value, mg KOH/g	0.1	0.05	0.01	0.01	0.1	0.1
11	Anticorrosive qualities–CETOP R 48H						
	Steel	–	0.3	–	–	–	–
	Copper	–	0.5	–	–	–	–
	Brass	–	1.9	–	–	–	–
	Zinc	–	2.2	–	–	–	–
	Aluminum	–	0.3	–	–	–	–
12	Fire resistance						
	Autoignition temperature–ASTM D 2155	545	610	620	485	545	585
	Spray ignition –7th Lux. report, para 3.1.2	6	3 max	Pass	Pass	Pass	6
	Wick test–RP 66H	9	5 max	Pass	Pass	Pass	9
	Flash point °C open Cup	245	270	274	243	245	265
	closed Cup	–	237	258	228	229	–
	Fire point °C	335	350	346	288	335	345
13	Seal compatibility index–ISO 6072	Butyl, EP, and PTFE	EP or fluorocarbon	Viton A, butyl, silicone, and EP			Viton, butyl
14	Effect on paint/recommended type	Epoxy paint	Epoxy paint	Reservoir should not be painted			Epoxy (cured)
15	Specific heat kJ/kg °C @ 25°C and 1,013 m bar	1.6	1.6	1.59	1.57	1.59	1.6
16	Coefficient of thermal expansion (volumetric)/°C	0.00069	0.00069	0.00069	0.00069	0.00069	0.00069
17	Thermal conductivity kJ/h/m/°C	–	–	–	–	–	0.46
18	Water separation–ASTM D 1401, minutes	–	5	–	–	–	5

Table A2–8 Phosphate Esters HFD-R

SI. No	Manufacturer / Brand	FUCHS Renosafe SF-D 0302M	GULF FR Fluid PE 46	Houghton Vaughan Houghton-Safe 1120	MOBIL Pyrogard 53	SHELL SFR D46	SMALL MAN Crown Press 2305
	Color	–	Clear, bright white/straw	Green	Blue	Clear green	Slightly hazy
1	Viscosity grade	46	46	46	46	46	46
2	Viscosity kinematic cSt @ 20°C	170	170	150	160	146	–
	40°C	42	42	43	43.7	43	43
	60°C	5.1 @ 100°C	18	17.5	18	17	–
3	Viscosity index	60	0	20	<0	20	–
4	Temperature in °C for viscosity of 850 cSt	0	4	5	6	6	–
5	Maximum operating bulk fluid temperature °C	<150	150	150	65	150	–
6	Relative density @ 15/15°C	1.140	1.13	1.126	1.15	1.125	1.125
7	Pour point °C	–18	–18	–15	–24	–18	–18
8	Foaming tendency/stability, mL	25/nil	20/nil	25/nil	20/nil	25/nil	–
9	Air release–minutes to 0.2% air @ 50°C	<2	15	–	6	12	–
10	Neutralization value, mg KOH/g	0.06	0.07	0.1	0.1	0.1 max	–
11	Anticorrosive qualities–CETOP R 48H						
	Steel	Per VDMA 24 317 and DIN 51515-2	Pass	Pass	Pass	–	–
	Copper		Pass	Pass	Pass	–	–
	Brass		Pass	Pass	Pass	–	–
	Zinc		Pass	Pass	Pass	–	–
	Aluminum		Pass		Pass		
12	Fire resistance						
	Autoignition temperature–ASTM D 2155	Per VDMA 24317 and DIN 51515-2	545	535	555	545	545
	Spray ignition–7th Lux. Report, para 3.1.2		Pass	Pass	Pass	<10	6
	Wick test–RP 66H		Pass	Pass	Pass	<10	9
	Flash point °C open cup		245	245	265	245	245
	closed cup		230				
	Fire point °C		335	340	308	335	335
13	Seal compatibility index–ISO 6072	Report	–	Chart available	Viton	Viton	–
14	Effect on paint/recommended type	Report	Epoxy	Chart	Epoxy	Epoxy	–
15	Specific heat kJ/kg °C @ 25°C and 1,013 m bar	3.3 @ 20°C	1.6	1.6	1.59	1.6	1.6
16	Coefficient of thermal expansion (volumetric)/°C	0.0007	–	0.00069	0.00069	0.00069	0.00069
17	Thermal conductivity kJ/h/m/°C	0.3-0.4	0.46	0.5	0.46	0.46	0.46
18	Water separation–ASTM D 1401, minutes	<3	5	–	20	30	–

Table A2-9 Phosphate Esters HFD-R

Sl. No.	Manufacturer / Brand	Reolube HYD 22	Reolube HYD 32	Reolube HYD 46	Reolube HYD 48	Reolube HYD 100	Reolube Turbofluid 46 × C
	FMC CORPORATION						
	Color	Clear liquid, no visible contamination					
1	Viscosity grade	22	32	46	68	100	46
2	Viscosity kinematic cSt @ 0°C	360	990	1,600	7,600	14,400	1,700
	20°C	–	–	–	–	–	175
	40°C	21.8	32.3	43	66	103	43.4
	100°C	3.8	4.7	5.3	5.7	6.2	5
3	Viscosity index (estimated)	–	–	–	–	–	–
4	Temperature in °C for viscosity of 850 cSt	–8	0	6	13	22	5
5	Maximum operating bulk fluid temperature °C	–20°C to 70°C	–20°C to 70°C	–20°C to 70°C	–20°C to 70°C	–20°C to 70°C	–20°C to 70°C
6	Density @ 15°C	1.18	1.15	1.125	1.139	1.146	1.14
7	Pour point °C	–35	–22	–18	–11	–8	–20
8	Foaming tendency/stability, mL @ Seq. 1	20/nil	20/nil	20/nil	20/nil	20/nil	30/nil
	Seq. 2	20/nil	20/nil	20/nil	20/nil	20/nil	10/0
	Seq. 3	20/nil	20/nil	20/nil	20/nil	20/nil	20/0
9	Air release–minutes to 0.2% air @ 50°C	–	–	8	–	–	1
10	Neutralization value mg KOH/g	0.06	0.06	0.06	0.06	0.06	0.06
11	Anticorrosive qualities–CETOP R48H						
	Steel	Pass	Pass	Pass	Pass	Pass	Pass
	Copper	Pass	Pass	Pass	Pass	Pass	Pass
	Brass	Pass	Pass	Pass	Pass	Pass	Pass
	Zinc	Pass	Pass	Pass	Pass	Pass	Pass
	Aluminum	Pass	Pass	Pass	Pass	Pass	Pass
12	Fire resistance						
	Autoignition temperature-ASTM D 2155	585	570	535	565	515	575
	Spray ignition–7th Lux. Report 3.1.2	Pass	Pass	Pass	Pass	Pass	3
	Wick test–ISO14935	Pass	Pass	Pass	Pass	Pass	5
	Flash point °C open cup	238	240	245	248	252	260
	Fire point °C	330	330	335	338	335	368
13	Seal compatibility	Butyl, teflon, viton, and EPDM					
14	Effect on paint/recommended type	Epoxy (cured)					
15	Specific heat kJ/kg °C @ 25°C and 1,013 m bar	1.5	1.5	1.6	1.6	1.6	1.6
16	Coefficient of thermal expansion (volumetric)/°C	0.00069	0.00069	0.00069	0.00069	0.00069	0.00069
17	Thermal conductivity kJ/hr/m/°C	–	–	0.46	0.46	–	0.46
18	Water separation–ASTM D 1401, minutes	–	–	–	–	–	1

Note: HYD fluids used for general industrial applications; 46XC used for turbine application

Appendix 3

DATA SHEETS ON BIODEGRADABLE HYDRAULIC FLUIDS

ACKNOWLEDGMENT

The data sheets given are extracts from the *BFPA Report-P67/ 1996*. Permission granted by the British Fluid Power Association is gratefully acknowledged. Data on Fuchs were provided by Fuchs Petrolube AG

Note: Data given are per British Standard unless otherwise specified and values given are only typical. Some companies are in the process of upgrading and rationalizing their products. For the current data or additional data, readers may contact the concerned fluid suppliers.

Table A3–1 Triglycerides–HETG

Sl. No.	Brand	BP OIL Biohyd 32	BP OIL Biohyd 46	CASTROL Carelube HTG 32	ESSO Hydraulic Oil PFL	MOBIL EAL 224 H
1	Viscosity grade	32	46	37	32/46	–
2	Viscosity kinematic cSt @ 0°C	–	–	190	200	246
	40°C	36	44	37	36	37.6
	100°C	8	10	8.2	8.4	8.54
3	Viscosity index	210	210	200	210	215
4	Temperature °C at which oil has a kinematic viscosity between 800 and 1,000 cSt	–20	–20	–22	–24	–23
5	Relative density 15/15°C	0.922	0.921	0.921	0.920	0.921
6	Pour point °C	–33	–33	–30	–27	–34
7	Flash point °C COC/closed cup	255/–	325/–	–/220	300/–	294/–
8	Water separation–ASTM D 1401, minutes	–	–	20	30	20
9	Foaming tendency/stability, mL Seq. 1	30/nil	50/nil	40/nil	–	nil/nil
	Seq. 2	30/nil	50/nil	100/trace	–	nil/nil
	Seq. 3	30/nil	50/nil	50/trace	–	nil/nil
10	Air release–minutes to 0.2% air @ 50°C	4	5	8	2	–
11	Neutralization value, mg KOH/g	0.9	0.75	1.2	0.22	–
12	Copper corrosion–3 hours @ 100°C	1A	1A	1	1	2A
13	Rust prevention–Procedure A distilled water	No rust	No rust	No rust	–	No rust
	Procedure B salt water	–	–	No rust	–	No rust
14	Acidity change after oxidation @ 1,000 hours— ASTM D 943, mg KOH/g	–	–	–	–	–
15	Seal compatibility index–ISO 6072	–	–	7	–	–
16	Vane pump wear test– total ring, and vane weight loss, mg	<150	<150	5	–	<50
17	Biodegradability–CEC-L-33-A-93, %	97	97	92	>96	>95
18	Baader oxidation test @ 95°C, 72 hours–DIN 51554 Part 3, % viscosity increase @ 40°C	<20	<20	–	–	10
19	FZG damaged load stage–DIN 51354 Part 2	>12	>12	12	12	12
20	Low temperature stability @ 72 hours— ASTM D 2532-87	1070	1200	–	–	–

212

Table A3-2 Triglycerides—HETG

Sl. No.		MORRIS	SHELL	SILKOLENE	TEXACO	
	Manufacturer					
	Brand	Liquimatic BVG 32	Naturelle HF-R 32	Planto-Hyd 40 N	Biostar Hydraulic 32	Biostar Hydraulic 46
1	Viscosity grade	32	32	32/46	32	46
2	Viscosity kinematic cSt @ 0°C	130	190	–	207	341
	40°C	31.9	35	40	32.63	46.2
	100°C	7.4	8.1	8.7	7.53	10.1
3	Viscosity index	205	215	210	210	210
4	Temperature °C at which oil has a kinematic viscosity between 800 and 1,000 cSt	–28	–24	–	–	–
5	Relative density 15/15°C	0.923	0.925	0.925	0.918	0.925
6	Pour point °C	–30	–36	–30	–30	–30
7	Flash point °C COC/closed cup	–/160	198/186	>250/–	210/–	225/–
8	Water separation–ASTM D 1401, minutes,	3	20	–	40	40
9	Foaming tendency/stability, mL Seq. 1	25/nil	100/nil	–	nil/nil	nil/nil
	Seq. 2	50/nil	130/nil	–	nil/nil	nil/nil
	Seq. 3	25/nil	110/nil	–	nil/nil	nil/nil
10	Air release–minutes to 0.2% air @ 50°C	10	7	–	5	7
11	Neutralization value, mg KOH/g	0.7	0.8	–	0.5	0.5
12	Copper corrosion –3 hours @ 100°C	1A	1	–	1B	1B
13	Rust prevention–Procedure A distilled water	No rust	Pass	–	–	–
	Procedure B salt water	No rust	Pass	–	–	–
14	Acidity change after oxidation @ 1,000 hours–ASTM D 943, mg KOH/g	<2	–	–	–	–
15	Seal compatibility index–ISO 6072	5	7	–	–	–
16	Vane pump wear test–total ring, and vane weight loss, mg	<50	<40	–	Pass	Pass
17	Biodegradability–CEC-L-33-A-93, %	>90	98	–	–	–
18	Baader oxidation test @ 95°C, 72 hours–DIN 51554 Part 3, % viscosity increase @ 40°C	–	–	–	Pass	–
19	FZG damaged load stage–DIN 51354 Part 2	10	12	–	10	12
20	Low temperature stability @ 72 hours–ASTM D 2532-87	–	–	–	12	12

Table A3-3 Triglycerides—HETG

Sl. No.	Brand	Manufacturer FUCHS		
		Plantohyd N 32	Plantohyd N 46	Plantohyd N 68
1	Viscosity grade	ISO VG 32	ISO VG 46	ISO VG 68
2	Viscosity kinematic cSt @ 0°C	160	300	450
	40°C	31.6	48.6	69.8
	100°C	7.4	10.6	13.9
3	Viscosity index	213	215	207
4	Temperature °C at which oil has a kinematic viscosity between 800 and 1,000 cSt	−26	−16	−10
5	Relative density 15/15°C	0.921	0.924	0.927
6	Pour point °C	−39	−36	−36
7	Flash point °C COC/closed cup	>270/−	>270/−	>270/−
8	Water separation—ASTM D 1401, minutes	−	−	−
9	Foaming tendency/stability, mL Seq. 1	20/0	30/0	20/0
	Seq. 2	15/0	20/0	15/0
	Seq. 3	10/0	10/0	10/0
10	Air release-minutes to 0.2% air @ 50°C	<7	<10	<10
11	Neutralization value, mg KOH/g	0.6	0.6	3.0
12	Copper corrosion −3 hours @ 100°C	1A-100 A 3	1A-100 A 3	1A-100 A 3
13	Rust prevention-Procedure A distilled water Procedure B salt water	0-A	0-A	0-A
14	Acidity change after oxidation @ 1, 000 hours—ASTM D 943, mg KOH/g	N A	N A	N A
15	Seal compatibility index-ISO 6072	Report	Report	Report
16	Vane pump wear test-total ring, and vane weight loss, mg	<150	<150	<150
17	Biodegradability-CEC-L-33-A-93, %	>90	>90	>90
18	Baader oxidation test @ 95°C, 72 hours-DIN 51554 Part 3, % viscosity increase @ 40°C	<20	<20	<20
19	FZG damaged load stage-DIN 51354 Part 2	>12	>12	>12
20	Low temperature stability at 72 hours—ASTM D 2532-87	Stable	Stable (−25°C)	Stable

N A = Not applicable

Table A3–4 Synthetic Esters-HEES

SI. No.	Manufacturer / Brand	BP OIL Biohyd SE 46	BP OIL Biohyd SE 68	ESSO Univis Bio SHP	MOBIL EAL Syndraulic 46	SHELL Naturelle HF-E 46
1	Viscosity grade	46	68	46	46	46
2	Viscosity kinematic cSt @ 0°C	–	–	385	378	390
	40°C	46	72	46.1	44.7	46
	100°C	9	13	8.4	8.01	9.1
3	Viscosity index	186	180	160	153	185
4	Temperature °C at which oil has a kinematic viscosity between 800 and 1,000 cSt	–20	–25	–11	–12	–15
5	Relative density 15/15°C	0.921	0.924	0.950	0.910	0.925
6	Pour point °C	–45	–42	–51	–42	–45
7	Flash point °C COC/closed cup	290/–	290/–	257/–	260/–	230/215
8	Water separation–ASTM D 1401, minutes	–	–	20	15	20
9	Foaming tendency/stability, mL Seq. 1	50/nil	50/nil	nil/nil	10/nil	30/nil
	Seq. 2	50/nil	50/nil	nil/nil	10/nil	20/nil
	Seq. 3	50/nil	50/nil	nil/nil	10/nil	35/nil
10	Air release– minutes to 0.2% air @ 50°C	3	4	6	–	8
11	Neutralization value, mg KOH/g	2.3	2.3	0.9	0.92	1.8
12	Copper corrosion–3 hours @ 100°C	1A	1A	1	1A	1
13	Rust prevention–Procedure A distilled water	No rust	No rust	Pass	No rust	Pass
	Procedure B salt water	–	–	Pass	No rust	Pass
14	Acidity change after oxidation @ 1,000 hours– ASTM D 943, mg KOH/g	–	–	–	–	–
15	Seal compatibility index–ISO 6072	–	–	–	–	8
16	Vane pump wear test–total ring, and vane weight loss, mg	<150	<150	<50	<50	<40
17	Biodegradability–CEC-L-33-A-93, %	>85	>85	>90	>95	90
18	Baader oxidation test @ 95°C, 72 hours– DIN 51554 Part 3, % viscosity increase @ 40°C	<20	<20	0	10	5.1 @ 110°C
19	FZG damaged load stage–DIN 51354 Part 2	>12	>12	12	12	>12
20	Low temperature stability at 72 hours– ASTM D 2532-87	–	–	–	–	–

215

Table A3-5   Synthetic Esters–HEES

Manufacturer		MORRIS LUBRICANTS			TEXACO		TOTAL OIL GREAT BRITAIN
Sl. No	**Brand**	Liquimatic BVG 46	Liquimatic BVG 68	Liquimatic BVG 100	Synstar Hydraulic 32	Synstar Hydraulic 46	Hydro Bio 46
1	Viscosity grade	46	68	100	32	46	46
2	Viscosity kinematic cSt @ 0°C	240	450	600	218	321	350
	40°C	45	67	99.7	32.5	45.2	47
	100°C	9.7	13.6	21.6	7.4	9.29	9.3
3	Viscosity index	205	215	245	194	194	185
4	Temperature °C at which oil has a kinematic viscosity between 800 and 1,000 cSt	−18	−15	−11	–	–	−10
5	Relative density 15/15°C	0.921	0.924	0.925	0.915	0.922	0.924
6	Pour point °C	−25	−25	−20	−50	−45	−42
7	Flash point °C COC/closed cup	–/190	–/200	–/200	215/–	252/–	242/–
8	Water separation–ASTM D 1401, minutes	3	4	5	30	30	25
9	Foaming tendency/stability, mL Seq. 1	25/nil	25/nil	40/nil	nil/nil	nil/nil	20/nil
	Seq. 2	50/nil	50/nil	60/nil	nil/nil	nil/nil	10/nil
	Seq. 3	25/nil	25/nil	40/nil	nil/nil	nil/nil	30/nil
10	Air release–minutes to 0.2% air @ 50°C	10	18	20	4	5	6
11	Neutralization value, mg KOH/g	0.7	0.9	0.9	–	–	0.2
12	Copper corrosion –3 hours @ 100°C	1A	1A	1A	1A	1A	1B
13	Rust prevention–Procedure A distilled water	No rust	No rust	No rust	–	–	Pass
	Procedure B salt water	No rust	No rust	No rust	–	–	Pass
14	Acidity change after oxidation @ 1,000 hours– ASTM D 943, mg KOH/g	–	–	–	–	–	–
15	Seal compatibility index–ISO 6072	<2	<2	<2	–	–	–
16	Vane pump wear test–total ring, and vane weight loss, mg	5	3	3	–	–	–
17	Biodegradability–CEC-L-33-A-93, %	<50	<50	<50	Pass	Pass	13.4
18	Baader oxidation test @ 95°C, 72 hours–DIN 51554 Part 3, % viscosity increase @ 40°C	>90	>90	>90	–	–	>90
19	FZG damaged load stage–DIN 51354 Part 2	11	11	11	6	4	12
20	Low temperature stability at 72 hours— ASTM D 2532-87	–	–	–	12	12	–

Table A3–6   Synthetic Esters–HEES

Sl. No	Manufacturer	FUCHS		
	Brand	Plantohyd 32 S	Plantohyd 46 S	Plantohyd 68 S
1	Viscosity grade	ISO VG 32	ISO VG 46	ISO VG 68
2	Viscosity kinematic cSt @ 0°C	200	350	500
	40°C	32	47.4	69
	100°C	7.1	9.3	13.0
3	Viscosity index	188	184	191
4	Temperature °C at which oil has a kinematic vscosity between 800 and 1, 000 cSt	−22	−15	−10
5	Relative density 15/15°C	921	921	923
6	Pour point °C	−39	−42	−39
7	Flash point °C COC/closed cup	246	290	304
8	Water separation–ASTM D 1401, minutes	<15	<15	<15
9	Foaming tendency/stability, mL Seq. 1	20/0	20/0	20/0
	Seq. 2	40/0	40/0	40/0
	Seq. 3	10/0	10/0	10/0
10	Air release–minutes to 0.2% air @ 50°C	5	7	9
11	Neutralization value–mg KOH/g	1.2	1.2	1.8
12	Copper corrosion–3 hours @ 100°C	1-100 A 3	1-100 A 3	1-100 A 3
13	Rust prevention–Procedure A distilled water	0-A	0-A	0-A
	Procedure B salt water	–	–	–
14	Acidity change after oxidation @ 1,000 hours–ASTM D 943, mg KOH/g	N A	N A	N A
15	Seal compatibility index–ISO 6072	Report	Report	Report
16	Vane pump wear test–total ring, and vane weight loss, mg	<150	<150	<150
17	Biodegradability–CEC-L-33-A-93, %	>90	>90	>90
18	Baader oxidation test @ 95°C, 72 hours–DIN 51554 Part 3, % viscosity increase @ 40°C	<20	<20	<20
19	FZG damaged load stage–DIN 51354 Part 2	>12	>12	>12
20	Low temperature stability at 72 hours–ASTM D 2532-87	Stable	Stable	Stable

NA = Not applicable.

Appendix 4

ISO CLASSIFICATION OF HYDRAULIC FLUIDS

Table A–4   ISO 6743-4: 1999, Lubricants, industrial oils and related products (class L)—Classification–Part 4: Family H (Hydraulic systems)

Code letter	General applications	Particular applications	More Specific applications	Compositions and properties	Symbol ISO-L	Typical applications	Remarks
H	Hydraulic systems	Hydrostatic		Non-inhibited refined mineral oils	HH		
				Refined mineral oils with improved antirust and antioxidation properties.	HL		
				Oils of HL type with improved antiwear performance	HM	General hydraulic systems, which include highly loaded components	
				Oils of HL type with improved viscosity/ temperature properties	HR		
				Oils of HM type with improved viscosity/ temperature properties	HV	Construction and marine equipment	
				Synthetic fluids with no specific fire resistant properties	HS		Special properties
			Applications where environmentally acceptable fluids are requested	Triglycerides Polyglycols Synthetic esters Polyalphaolefin and related hydrocarbon products	HETG HEPG HEES HEPR	General hydraulic systems (mobile)	The minimum content of base fluid for each category shall not be less than 70% (m/m)
			Hydraulic slideway systems	Oils of HM type with anti stick/slip properties	HG	Machines with combined hydraulic and plain bearing way lubrication systems where vibration or intermittent sliding (stick/slip) at low speed is to be minimized	These fluids are intended to be multi-functional, but do not function successfully under all hydraulic applications

Applications where fire resistant fluids are required	Oil-in-water emulsions	HFAE	Typically more than 80% mass fraction of water
	Chemical solutions in water	HFAS	Typically more than 80% mass fraction of water
	Water-in-oil emulsions	HFB	
	Water polymer solutions	HFC*	Typically more than 35 % mass fraction of water
	Synthetic fluids containing no water and consisting of phosphate esters	HFDR*	
	Synthetic fluids containing no water and of other compositions	HFDU*	
Hydrokinetic	Automatic transmissions	HA	
	Couplers and converters	HN	Classification concerning those applications has not been examined in detail and can be supplemented

*Fluids of this type may also fulfill the requirements of eco-biodegradability and eco-toxicity defined in HE categories

Appendix 5

ISO CLASSIFICATION OF VISCOSITY GRADES

Table A–5   ISO 3448:1992, Industrial Liquid Lubricants–ISO Viscosity Classification

ISO viscosity grade	Midpoint viscosity cSt @ 40°C (mm²/sec)	Kinematic viscosity limits at 40°C	
		Minimum	Maximum
ISO VG 2	2.2	1.98	2.42
ISO VG 3	3.2	2.88	3.52
ISO VG 5	4.6	4.14	5.06
ISO VG 7	6.8	6.12	7.48
ISO VG 10	10	9.00	11.00
ISO VG 15	15	13.50	16.50
ISO VG 22	22	19.80	24.20
ISO VG 32	32	28.80	35.20
ISO VG 46	46	41.40	50.60
ISO VG 68	68	61.20	74.80
ISO VG 100	100	90.00	110.00
ISO VG 150	150	135.00	165.00
ISO VG 220	220	198.00	242.00
ISO VG 320	320	288.00	352.00
ISO VG 460	460	414.00	506.00
ISO VG 680	680	612.00	748.00
ISO VG 1,000	1,000	900.00	1,100.00
ISO VG 1,500	1,500	1,350.00	1,650.00

Appendix 6

VISCOSITY CONVERSION FROM cSt TO SUS, ENGLER AND REDWOOD

Table A–6   Viscosity: Approximate Equivalent at Same Temperature

Kinematic viscosity cSt	Saybolt universal seconds	Redwood No. 1 seconds	Engler degrees	Saybolt furol seconds	Redwood No. 2 seconds
1.8	32	30.8	1.14	–	–
2.7	35	32.2	1.18	–	–
4.2	40	36.2	1.32	–	–
5.8	45	40.6	1.46	–	–
7.4	50	44.9	1.60	–	–
8.9	55	49.1	1.75		
10.3	60	53.5	1.88	–	–
11.7	65	57.9	2.02	–	–
13.0	70	62.3	2.15	–	–
14.3	75	67.6	2.31	–	–
15.6	80	71.0	2.42	–	–
16.8	85	75.1	2.55	–	–
18.1	90	79.6	2.68	–	–
19.2	95	84.2	2.81	–	–
20.4	100	88.4	2.95	–	–
22.8	110	97.1	3.21	–	–
25.0	120	105.9	3.49	–	–
27.4	130	114.8	3.77	–	–
29.6	140	123.6	4.04	–	–
31.8	150	132.4	4.32	–	–
34.0	160	141.1	4.59	–	–
36.0	170	150.0	4.88	–	–
38.4	180	158.8	5.15	–	–
40.6	190	167.5	5.44	–	–
42.8	200	176.4	5.72	23.0	–
47.2	220	194.0	6.28	25.3	–

(continues)

Table A–6 *(Continued)*

Kinematic viscosity cSt	Saybolt universal seconds	Redwood No. 1 seconds	Engler degrees	Saybolt furol seconds	Redwood No. 2 seconds
51.6	240	212	6.85	27.0	–
55.9	260	229	7.38	28.7	–
60.2	280	247	7.95	30.5	–
64.5	300	265	8.51	32.5	–
69.9	325	287	9.24	35.0	–
75.3	350	309	9.95	37.2	–
80.7	375	331	10.70	39.5	–
86.1	400	353	11.40	42.0	–
91.5	425	375	12.10	44.2	–
96.8	450	397	12.80	47.0	–
102.2	475	419	13.5	49	–
107.6	500	441	14.2	51	–
118.4	550	485	15.6	56	–
129.2	600	529	17.0	61	–
140.3	650	573	18.5	66	–
151	700	617	19.9	71	–
162	750	661	21.3	76	–
173	800	705	22.7	81	–
183	850	749	24.2	86	–
194	900	793	25.6	91	–
205	950	837	27.0	96	–
215	1,000	882	28.4	100	–
259	1,200	1,058	34.1	121	104
302	1,400	1,234	39.8	141	122
345	1,600	1,411	45.5	160	138
388	1,800	1,587	51	180	153
432	2,000	1,763	57	200	170
541	2,500	2,204	71	250	215
650	3,000	2,646	85	300	255
758	3,500	3,087	99	350	300
866	4,000	3,526	114	400	345
974	4,500	3,967	128	450	390
1,082	5,000	4,408	142	500	435
1,190	5,500	4,849	156	550	475
1,300	6,000	5,290	170	600	515
1,405	6,500	5,730	185	650	560
1,515	7,000	6,171	199	700	600
1,625	7,500	6,612	213	750	645
1,730	8,000	7,053	227	800	690
1,840	8,500	7,494	242	850	730
1,950	9,000	7,934	256	900	770
2,055	9,500	8,375	270	950	815
2,165	10,000	8,816	284	1,000	855

(Courtesy: ExxonMobil Lubricants and Petroleum Specialties, Houston, Texas, U.S.)

MEASUREMENT CONVERSION TABLES

Table A–7 Measurement Conversion Tables

Physical quantity	Unit to be converted		Multiply by	Converted unit	
Density (liquids)	gram/cubic centimeter	g/cm^3	0.0361	pound/cubic inch	lb/in^3
	kilogram/cubic meter	kg/m^3	0.008353	pound/U.S. gallon	lb/gal
Density (solids)	kilogram/cubic meter	kg/m^3	0.062428	pound/cubic foot	lb/ft^3
Force	dyne	dyn	1×10^{-5}	Newton	N
	kilogram	kgf	9.80665	Newton	N
	Newton	N	0.2248	pound force	lbf
	pound force	lbf	4.448	Newton	N
Length	foot	ft	0.3048	meter	m
	inch	in	2.54	centimeter	cm
	meter	m	3.2808	foot	ft
	millimeter	mm	0.0394	inch	in
	microinch	μin	0.0254	micron	μm
	micron	μm	39.37	microinch	μin
Mass	gram	g	0.03527	ounce	oz
	kilogram	kg	2.2046	pound	lb
	metric ton (tonne)	mt	1000	kilogram	kg
	pound	lb	0.4536	kilogram	kg
Power	British thermal unit	btu	1.0551	kilojoules	kJ
	btu/hour	btu/h	0.2931	watt	W
	Horsepower	hp	745.7	watt	W
	kilocalories/hour	kcal/h	1.163	watt	W
	kilowatt hour	kWh	3.6	megajoule	MJ
	metric horsepower	PS	735.5	watt	W
	watt	W	1.34×10^{-3}	Horsepower	hp

Quantity	Unit		Factor	Unit	
Pressure	atmosphere	Atm	1.01325	bar	bar
	bar	bar	1×10^5	Newton/m²	N/m²
	bar	bar	100	kilopascal	kPa
	bar	bar	14.5038	pound/m²	psi
	foot of water (17°C)	ft H$_2$O	0.02946	atmosphere	atm
	inches of mercury(60°F)	in.Hg	33.7684	millibar	mbar
	Pascal	Pa	1.0	Newton/m²	N/m²
	pound/square inch	psi	6894.757	Newton/m²	N/m²
	pound/square inch	psi	0.0689457	bar	bar
Specific heat	btu/pound °F	btu/lb °F	4186.8	Joule/kg Kelvin	J/kg K
	Joule/kg Kelvin	J/kg K	2.388×10^{-4}	btu/pound °F	btu/°F
Surface Tension	dyne/centimeter	dyn/cm	1.0	milli Newton/meter	mN/m
	pound force/foot	lbf/ft	14.594	Newton/meter	N/m
Temperature	degree Celsius	°C	9/5°C+32	Fahrenheit	°F
	degree Celsius	°C	°C+273.16	Kelvin	K
	degree Fahrenheit	°F	5/9(°F−32)	Celsius	°C
Thermal conductivity	btu/hour foot °F	btu/h	1.7307	watt/meter Kelvin	w/m.K
	watt/meter Kelvin	w/m. K	0.578	btu/hour foot °F	btu/h ft °F
Torque	Newton meter	Nm	8.8507	pound force inch	lbf in
	pound force inch	lbf in	0.113	Newton meter	Nm
Velocity	foot/second	ft/sec	0.3048	meter/second	m/sec
	meter/second	m/sec	3.2808	foot/second	ft/sec
	meter/minute	m/min	0.05468	foot/second	ft/sec
	millimeter/second	mm/sec	0.197	foot/minute	ft/min
	revolutions/minute	rpm	0.10472	radians/second	rad/sec

(continues)

Table A-7 (Continued)

Physical quantity	Unit to be converted		Multiply by	Converted unit	
Viscosity	centipoise	cP	1.0	milli Newton sec/m^2	mNsec/m^2
	centipoise	cP	6.72×10^{-4}	pound/second foot	lb/s ft
	centistokes	cSt	1×10^{-6}	square meter/second	m^2/sec
	centistokes	cSt	10.76×10^{-6}	square foot/second	ft^2/sec
Volume (liquid)	barrel (petroleum oil)	bbl	42.0	U.S. gallon	gal
	liter	L	0.264	U.S. gallon	gal
	liter	L	1000	cubic centimeter	cm^3
	liter	L	0.03531	cubic foot	ft^3
	U.S. gallon	gal	3.7854	liter	L
	U.S. gallon	gal	3.7854×10^{-3}	cubic meter	m^3
Volumetric flow rate	cubic feet/minute	ft^3/min	0.47195	liters/second	L/sec
	cubic meter/second	m^3/s	1.585×10^4	U.S. gallons/minute	gal/min
	gallon/minute	gal/min	63.09	cubic centimeter/sec	cm^3/sec
	liter/second	L/sec	15.85	U.S. gallons/minute	gal/min

REFERENCES

CHAPTER 1

1. McNeil, I, *Hydraulic Power*, Longmans, London, 1972.
2. *Water Hydraulics: A Technical Guide*, BFPA Report No. P82-1999, The British Fluid Power Association, Chipping Norton, U.K.
3. *Water Hydraulics: A Guide to Products and Suppliers in the U.K.*, BFPA Report No. P75-1997, ibid.
4. Usher, S., "Water Hydraulic Power: The Technology Matures", 47th *National Conference on Fluid Power*, National Fluid Power Association, Milwaukee, WI, U.S., 1996.
5. Usher, S., "Water Hydraulic Extraction and Rescue Tools", *ibid*.
6. Usher, S., "Water Hydraulic Application Examples," *International Workshop on Water Hydraulics*, Copenhagen, Denmark, 1999.
7. Usher, S. and Young, M., "A Decade of Water Hydraulics," *Water Hydraulics Conference*, Tampere, Finland, 1997.
8. *Why water hydraulics? And What is the difference?* 1995, Hauhinco Water Hydraulics Ltd., Heckmondwike, U.K.
9. *Ecologically Acceptable Hydraulic Fluids: Data Sheets*, BFPA Report No. P67-1996, The British Fluid Power Association, Chipping Norton, U.K.
10. Grover, K.B. and Perez, R.J., "The Evolution of Petroleum-Based Hydraulic Fluids, *Lubrication Engineering*, **46, 1,** Jan.1990, pp. 15–20.

CHAPTER 2

1. Speight, J., "Petroleum Refining," *Encyclopedia of Energy, Technology and the Environment*, Volume 3. Bisco, A. and Boots, S., Eds. John Wiley & Sons, Inc., U.S., 1995, pp. 2,237–2,259.
2. Speight, J., "Hydro Processing," *ibid*, pp. 1,791–1,795.
3. Schaefer-Pederson, B.; Thompson, G.J.; and Tippett, T.W.; "Hydrocracking," *ibid*, pp. 1,745–1,752.
4. Gilbert, J.B.; Kartzmark, R.; and Sproule, L.W.; "Hydrogen Process of Lube Stocks," *Journal of the Institute of Petroleum*, **53, 526,** Oct. 1967, pp. 317–327.

5. ISO 6743-4: 1999, *Lubricants, Industrial Oils and Related Products (Class L) Classification–Part 4; Family H (Hydraulic Systems)*, International Organization for Standardization, Geneva, Switzerland.

6. *Hydraulic Fluids Standards*, 1983, Denison Hydraulics France S.A., Vierzon Cedex, France.

7. DIN 51524: 1985, *Hydraulic Oils–Minimum Requirements, HL Hydraulic oils–Part 1; HLP Hydraulic Oils–Part 2; HLVP Hydraulic Oils–Part 3*; Deutsches Institut fur Normung e.V., Berlin, Germany.

8. *Data Sheets on Antiwear and R&O Hydraulic Oils Specifications*, 1998, Cincinnati Machine, Cincinnati, OH, U.S.

9. *Hydraulic Fluids, Mineral Oil: Data Sheets*, BFPA Report No. P12-1996, The British Fluid Power Association, Chipping Norton, U.K.

10. Synder, C.E., Jr.; Gschwender, L.J.; Paciorek, K.; Kratzer, R.; and Nakahara, J.; "Development of Shear Stable Viscosity-Index Improvers for Hydraulic Fluids," *Lubrication Engineering*, **42, 9**, Sep. 1986, pp. 547–557.

11. *Hydraulics–Fluids and Systems*, 1989, BP Oil International Ltd., London, U.K.

12. *Hydraulic Systems for Industrial Machines*, 1970, Mobil Publications, Mobil Oil Corporation, New York, U.S.

13. *Industrial Hydraulic Fluids*, 1995, Exxon Brochure DG-2C, Exxon Company, Houston, TX, U.S.

14. Hamm F.A., "Hydraulics II," *Lubrication*, **32, 2,** Caltex Petroleum Corporation, Irving, TX, U.S., 1982, pp. 13–24.

15. "Hydraulics," *Lubrication*, **82, 1,** Texaco Services (Europe) Ltd., Brussels, Belgium, 1996, pp. 1–24.

16. *Technical Information Sheets on Shell Hydraulic Fluids*, 1996, Shell U.K. Ltd., Manchester, U.K.

17. de la Flor, F.M. and Palmer, W.E., "Development of a New Heavy Duty Automatic Transmission Fluid C-4 Specifications," SAE Technical Paper No. 892160, *SAE Transactions, Section 4*, 1989, pp. 1,300–1,310.

18. Papay, A.G., "Automatic Transmission Fluids Dexron II and Beyond," *Lubrication Engineering*, **45, 2,** Feb. 1989, pp. 121–128.

19. Ashikawa, R.; Naruse, T.; Kurashina, H.; Matsuoka, T.; Adachi, T.; and Nakayama, T.; "ATF Characteristics Required for the Latest Automatic Transmissions," SAE Technical Paper No. 932849, *SAE Transactions, Section 4*, 1993, pp. 1,923–1,929.

20. Papay, A.G., "Effect of Chemistry on Performance of Automatic Transmission Fluids," *Lubrication Engineering*, Aug. 1990, **46, 8,** pp. 511–518.

21. Papay, A.G., "Formulating Automatic Transmission Fluids," *Lubrication Engineering*, **47, 4,** April 1991, pp. 271–275.

22. Cave, W., and Lochte, M., "Development of an Updated Brake Chatter Test for Anti-Brake Chatter Transmission/Hydraulic fluids," SAE Technical Paper No. 961817, *SAE Transactions, Section 2*, 1996, pp. 281–285.

23. *Automatic Transmission Fluids*, Technical Bulletins, Lubrizol Corporation, Wickliffe, OH, U.S.

24. ISO 3448: 1992, *Industrial Liquid Lubricants–ISO Viscosity Classification*, International Organization for Standardization, Geneva, Switzerland.

25. *Hydraulic Fluid and Temperature: Recommendations for Industrial Machinery*, Technical Information Data sheet 1-286-S, 1985, Vickers Inc. Troy, MI, U.S.

26. ISO 6073: 1997, *Petroleum Products–Prediction of Bulk Moduli of Petroleum Fluids Used in Hydraulic Fluid Power Systems*, International Organization for Standardization, Geneva, Switzerland.

27. Hayward, A.T.J., "The Compressibility of Hydraulic Fluids," *Journal of the Institute of Petroleum*, U.K., **51, 494,** Feb. 1965, pp. 35–52.

28. Okutsu, R.; Outa, E.; Kuramochi, S.; and Machiyama, T.; "Noise and Vibration Induced by Throttling of High-Pressure Hydraulic Fluids," *Bulletin of JSME*, **28, 239,** May 1985, pp. 837–844.

29. Leshchenko, V.A., et al., "Undissolved Gases in Hydraulic System Fluids," *Soviet Engineering Research*, **5, 11,** Nov. 1985, pp. 23–26.

30. Tsuji, S., and Katakura, H., "A Fundamental Study of Aeration in Oil, Second Report: The Effect of the Diffusion of Air on the Diameter Change of a Small Bubble Rising in a Hydraulic Oil," *Bulletin of JSME*, **21, 156,** June 1978, pp. 1015–1021.

31. Claxton, P.D., "Aeration of Petroleum-Based Steam Turbine Oils," *Tribology*, Feb. 1972, pp. 8–13.

32. Rowand, H.H., Jr.; Patula, R.J.; and Sargent, L.B., Jr.; "The Evaluation of the Air Entraining Tendency of Fluids," *Lubrication Engineering*, **29, 11,** Nov. 1973, pp. 491–497.

33. Volpato, G.A.; Manzi, A.G.; and Del Ross, S.; "A New Method to Study Air Entrainment in Lubricating Oils," *First European Tribology Conference*, London, 1973, pp. 335–342.

34. Hayward, A.T.J., "How to Avoid Aeration in Hydraulic Circuits," *Hydraulics and Pneumatics*, Nov. 1963, pp. 79–83.

35. Magorien, V.C., "How Hydraulic Fluids Generate Air," *Hydraulics and Pneumatics*, June 1968, pp. 104–108.

36. Magorien, V.C., "Effects of Air on Hydraulic Systems," *Hydraulics and Pneumatics*, Oct. 1967, pp. 128–131.

37. Totten, G.E.; Sun, Y.H.; and Bishop, R.J., Jr.; "Hydraulic Fluids: Foaming, Air Entrainment, and Air Release–A Review," SAE Technical Paper No. 972789, *SAE Transactions, Section 2*, 1997, pp. 379–389.

38. *Hydraulic Hints and Troubleshooting Guide*, 1996, Bulletin No. 694-1996, Vickers Inc., Troy, MI, U.S.

39. Galvin, G.D., *Chemistry of Lubricants*, M.Sc. Tribology Course Materials, 1974, University of Leeds, Leeds, U.K.

40. Summers-Smith, D., "The Selection of Lubricating Oils for Industrial Plant: The User's Point of View," *Proceedings of the Institution of Mechanical Engineers*, **187, 47/73,** 1973, pp. 493–500.

41. Papay, A.G., and Harstick, C.S., "Petroleum-Based Industrial Hydraulic Oils–Present and Future Developments," *Lubrication Engineering*, **31, 1,** Jan. 1975, pp. 6–15.

42. Grover, K.B., and Perez, R.J., "The Evolution of Petroleum-Based Hydraulic Fluids," *Lubrication Engineering*, **46, 1,** Jan. 1990, pp. 15–20.

43. Schmidt, R.H.; Poole, R.J.; and Shim, J.; "From Water to Super Stabilized Antiwear Hydraulic Oils," *SAE Technical Paper No. 780780*, Society of Automotive Engineers, Warrendale, PA, U.S., 1978.

44. Okada, M., and Yamashita, M., "Development of an Extended Life Hydraulic Fluid," *Lubrication Engineering*, **43, 6,** June 1987, pp. 459–466.

45. Forbes, E.S., "Antiwear and Extreme Pressure Additives for Lubricants," *Tribology,* **3, 3,** Aug. 1970, pp. 145–152.
46. Kotvis. P.V., "Overview of the Chemistry of Extreme Pressure Additives," *Lubrication Engineering,* **42, 6,** June 1986, pp. 363–366.
47. Soul, D., "The Development and Evaluation of Antiwear Additive Systems for Hydraulic Oils in Relation to Equipment Design," *Proceedings of the International Symposium on Performance Testing of Hydraulic Fluids,* Institute of Petroleum, London, 1978, pp. 133–148.
48. Spedding, C.N., and Walkins, R.C., "The Antiwear Mechanism of ZDTP–Part.1," *Tribology International,* **15, 1,** Jan. 1982, pp. 9–12.
49. Walkins, R.C., "The Antiwear Mechanism of ZDTP–Part.2," *Tribology International,* **15.1,** Jan. 1982, pp. 13–16.
50. Schmitt, R.H.; Armstrong, E.L.; Wooding, P.S.; and Baudouin, P.; "Performance Testing of New Generation Stabilized Antiwear Hydraulic Oils," *Proceedings of the International Symposium on Performance Testing of Hydraulic Oils,* Institute of Petroleum, London, 1978, pp. 116–131.
51. Ponjee, A.L.; Schmitt, R.H.; Yazawa, H.; and Yokota, N.; "Modern Hydraulic Fluids–Balanced Performance Testing," *Japan International Lubrication Conference,* Tokyo, 1975, pp. 650–658.
52. Saxena, D.; Mookken, R.T.; Srivastava, S.P.; and Bhatnagar, A.K.; "An Accelerated Aging Test for Antiwear Hydraulic Oils," *Lubrication Engineering,* **49, 10,** Oct. 1993, pp. 801–809.
53. Perez, R.J., and Brener, M.S., "Development of New Constant Volume Vane Pump Test for Measuring Wear Characteristics of Fluids," *Lubrication Engineering,* **48, 5,** May 1992, pp. 354–359.
54. Recchuite, A.D., and Newingham, T.D., "Effect of Zinc Dithiophosphate on Axial Piston Pump Performance," *Lubrication Engineering,* **32, 9,** Sep.1976, pp. 481–488.
55. Brown, C.L., "Bearing Bronze Wear in Hydraulic and Lubricating Environments," *Lubrication Engineering,* **28, 11,** Nov. 1972, pp. 408–411.
56. Fainman, M. Z., and Hiltner, L.G., "Compatibility of Elastomeric Seals and Fluids in Hydraulic Systems," *Lubrication Engineering,* **37, 3,** March 1981, pp. 132–137.
57. Murray, K., "Compatibility of Hydraulic System Materials," *Lubrication Engineering,* **32, 6,** June 1976, pp. 299–305.
58. Allen, J.M., and Robertson, R.S., "A Study of Oil Performance in Numerically Controlled Hydraulic Systems," *Proceedings of the National Conference on Fluid Power,* National Fluid Power Association. Milwaukee, WI. U.S., 1974, pp. 435–454.

CHAPTER 3

1. *Annual Book of ASTM Standards, 1995, Petroleum Products and Lubricants, Volumes 05.01, 05.02, and 05.03,* American Society for Testing and Materials, West Conshohocken, PA, U.S.
2. McCulloch, H.W., Jr., "Hydraulic Oil Tests You Should Know," *Hydraulics and Pneumatics,* Feb. 1978, pp. 55–59.

3. Burgoyne, J.H.; Roberts, A.F.; and Alexander, J.L.; "The Significance of Open Flash Points," *Journal of the Institute of Petroleum*, **53, 526,** Oct. 1967, pp. 338–341.

4. Kuchta, J.M., and Cato, R.J., "Ignition and Flammability Properties of Lubricants," SAE Technical Paper No. 680323, *SAE Transactions*, 1968, pp. 1,008–1,020.

5. Hayward, A.T.J., "The Viscosity of Bubbly Oil," *Journal of the Institute of Petroleum*, **48, 461,** 1962, pp. 156–164.

6. Claxton, P.D., "Aeration of Petroleum-Based Steam Turbine Oils," *Tribology*, Feb. 1972, pp. 8–13.

7. Rowand, H.H., "The Evaluation of the Air Entraining Tendency of Fluids," *Lubrication Engineering*, **29, 11,** Nov. 1973, pp. 491–497.

8. Volpato, G.A.; Manzi, A.G.; and Del Ross, S.; "A New Method to Study Air Entrainment in Lubricating oils," *Proceedings of the First European Tribology Conference*, London, 1973, pp. 335–342.

9. Totten, G.E.; Sun, Y.H.; and Bishop, R.J., Jr.; "Hydraulic Fluids: Foaming, Air Entrainment, and Air Release–A Review," SAE Technical Paper No. 972789, *SAE Transactions, Section 2*, 1997, pp. 379–389.

10. Reynard, R., and Dalibert, A., "On the Evaluation of Mechanical Properties of Hydraulic Oils," *Journal of the Institute of Petroleum*, **55, 542,** March 1969, pp. 109–116.

11. Soul, D.M., "The FZG Rig for Evaluating Industrial Lubricants," *Lubrication Engineering*, **31, 5,** May 1975, pp. 254–260.

12. Klaus, E.E., and Perez, J.M., "Comparative Evaluation of Several Hydraulic Fluids in Operational Equipment–a Full-Scale Pump Test Stand and the Four Ball Wear Tester," *SAE Technical Paper No. 831680*, Society of Automotive Engineers, Warrendale, PA, U.S., 1983.

13. Klaus, E.E., and Perez, J.M., "Comparative Evaluation of Several Hydraulic Fluids in Operational Equipment–a Full-Scale Pump Test Stand and the Four Ball Wear Tester, *Part II*–Phosphate Esters, Glycols and Mineral Oils," *Lubrication Engineering*, **46, 4,** April 1990, pp. 249–255.

14. Tessmann, R.K., and Hong, T., "An Effective Bench Test for Hydraulic Fluid Evaluation," SAE Technical Paper No. 932438, *SAE Transactions, Section 2*, 1993, pp. 406–415.

15. Perez, R.J., and Brenner, S.M., "Development of New Constant Volume Vane Pump Test for Measuring Wear Characteristics of Fluids," *Lubrication Engineering*, **48, 5,** May 1992, pp. 401–409.

16. Broszeit, E., and Steindorf, H., "Testing of Hydraulic Fluids with Large Vickers 35 VQ 25 Pump [in German]," *Oel Hydraulik und Pneumatik*, **32, 2,** Feb. 1988, pp. 101–105.

17. Gellrich, R.; Kunz, A.; Beckmann, G.; and Broszeit, E.; "Theoretical and Practical Aspects of the Wear of Vane Pumps, Part A: Adaptation of a Model for Predictive Wear Calculation," *Wear*, **181–183,** 1995, pp. 862–867.

18. Kunz, A.J.; Gellrich, R.; Beckmann, G.; and Broszeit, E.; "Theoretical and Practical Aspects of the Wear of Vane Pumps, Part B: Analysis of Wear Behavior in the Vickers Vane Pump Test," *Wear*, **181–183,** 1995, pp. 868–875.

19. Kunz, A.J., and Broszeit, E., "Comparison of Vane Pump Tests Using Different Vickers Vane Pumps," *Tribology of Hydraulic Pump Testing, ASTM STP-1310,*

American Society for Testing and Materials, West Conshohocken, PA, U.S., 1996.

20. Thoenes, H.W.; Bauer, K.; and Herman, P.; "Testing the Antiwear Characteristics of Hydraulic Fluids: Experience with Test Rigs Using a Vickers Pump," *Proceeding of the International Symposium on Performance Testing of Hydraulic Fluids*, Institute of Petroleum, London, 1978.

21. Schmitt, R.H.; Poole, R.J.; and Shim, J.; "From Water to Super Stabilized Antiwear Hydraulic Oils," *SAE Technical Paper No. 780780*, Society of Automotive Engineers, Warrendale, PA, U.S., 1978.

22. Reichel, J., "Mechanical Testing of Hydraulic Fluids," *Proceedings of the 11th International Colloquium*, Technische Akademie Esslingen, Germany, Jan. 1998, pp. 1,825–1,836.

23. Kessler, M., and Feldman, D.G., "Development of a New Application Related Test Procedure for Mechanical Testing of Hydraulic Fluids," *Proceeding of the 6th Scandinavian International Conference on Fluid Power, SICFP 1999*, Tampere, Finland.

24. Denison Standard T5D-042, *Procedure for Vane Pump Wear Test and Denison 46 series Axial Piston Pump Wear Test*, 2000, Denison Hydraulics Inc., Marysville, OH, U.S.

25. Denison Standard A-TP-30283, *Hydraulic Fluid Performance Evaluation on Vane Pumps*, May 2000, Denison Hydraulics France S.A., Vierzon Cedex, France

26. BS 2000-Part 281: 1983, *Antiwear Properties of Hydraulic Fluids by the Vane Pump Test*, British Standards Institution, London, U.K.

27. DIN 51389–Part 2: 1982, *Determination of Lubricants: Mechanical Testing of Hydraulic Fluids in the Vane-Cell Pump*, Deutsches Institut fur Normung e.V., Berlin, Germany.

28. *Hydraulics Hints and Troubleshooting Guide*, Bulletin No. 694-1996, Vickers Inc., Troy, MI, U.S.

29. DIN 51354-Part 2: 1990, *Testing of Lubricants: FZG Gear Test Rig Method A/8.3/90 for Lubricating Oils*, Deutsches Institut fur Normung e.V., Berlin, Germany.

30. IP 334/1993, *Determination of Load Carrying Capacity of Lubricants–FZG Gear Machine Method*, Institute of Petroleum, London, U.K.

31. DIN 51554-Part 3: 1978, *Testing of Mineral Oils: Testing of Aging According to Baader; Object Sampling, Aging*, Deutsches Institut fur Normung e.V., Berlin, Germany.

32. Denison Standard F-TP-02100-A, *Procedure for Determining Filterability of Hydraulic fluids*, Sep. 2001, Denison Hydraulics France S.A., Vierzon Cedex, France.

33. Litt, F.A., "Precision in Wet Filterability Determinations," *Lubrication Engineering*, **39, 7,** July 1983, pp. 448–449.

34. AFNOR NFE 48-690 and NFE 48-691, 1993, *Hydraulic Oil Filterability Test*, French National Standards Organization, Paris, France.

35. ISO 13357-2: 1998, *Petroleum Products–Determination of Filterability of Lubricating Oils, Part 2: Procedure for Dry Oils*, International Organization for Standardization, Geneva, Switzerland.

36. BS 4832-87, *Compatibility Between Elastomeric Materials and Hydraulic Fluids*, British Standards Institution, London, U.K.

37. ISO 6072: 1999, *Hydraulic Fluid Power–Compatibility Between Elastomeric Materials and Fluids*, International Organization for Standardization, Geneva, Switzerland.

38. DIN 53521: 1979, *Testing of Natural and Synthetic Rubbers, Determination of the Behavior Against Fluids, Moisture and Gases*, Deutsches Institut fur Normung e.V, Berlin, Germany.

39. Murray, K., "Compatibility of Hydraulic System Materials," *Lubrication Engineering*, **32, 6,** June 1976, pp. 299–305.

40. ASTM D 2070-91, "Standard Test Method for Thermal Stability of Hydraulic Oils," *Annual Book of ASTM Standards, Volume 05.01*, American Society for Testing and Materials, West Conshohocken, PA, U.S., 1995, pp. 681–684.

41. ASTM D 2160-92, "Standard Test Method for Thermal Stability of Hydraulic Fluids," *ibid*, pp. 697–700.

42. *Thermal Stability Test Procedure A and Procedure B*, Cincinnati Machine 10-SP-95046, Cincinnati, OH, U.S.

43. IP 294/1993, *Determination of Shear Stability of Polymer Containing Oils–Diesel Injector Rig Method*, Institute of Petroleum, London, U.K.

44. ASTM D 3945-93, "Standard Test Methods for Shear Stability of Polymer-Containing Fluids Using a Diesel Injector Nozzle," *Annual book of ASTM Standards, Volume 05.02*, American Society for Testing and Materials, West Conshohocken, PA, U.S., 1995, pp. 609–617.

45. ASTM D 5621-94, "Standard Test Method for Sonic Shear Stability of Hydraulic Fluid," *Annual Book of ASTM Standards, Volume 05.03*, American Society for Testing and Materials, West Conshohocken, PA, U.S., 1995, pp. 784–786.

46. Sarbach, R.A., Denison Hydraulics Inc., Marysville, OH, U.S. *Private Communication*, Dec. 2000.

47. *Hydraulic Fluid, Fire Resistant, Aqueous Polyglycol*, Interim Defense Standard 91-110/1, Dec.31, 1997, Joint Service Designation:OX-40, Ministry of Defense, U.K.

CHAPTER 4

1. Hoffmann, M.N., "Hydraulic Fluid of 95-percent Water?" *Lubrication Engineering* **35, 2,** Feb. 1979, pp. 65–71.

2. Adams, R.J., "How Ford Motor Company Tests High Water Base Hydraulic Fluids," *Hydraulics and Pneumatics, 34, 5,* May 1981, pp. 76–79.

3. "High Water Base Fluids: a Ford Better Idea," *Hydraulics and Pneumatics,* **34, 10,** Oct.1981, pp. 228–238.

4. Springborn, R.K. (Ed), *Cutting and Grinding Fluids: Selection and Application*, American Society of Tool and Manufacturing Engineering, 1966.

5. Reichel, J., "Fluid Power Engineering with Fire Resistant Hydraulic Fluids–Experience with Water-Containing Hydraulic Fluids," *Lubrication Engineering, 50, 12,* Dec. 1994, pp. 947–952.

6. *Technical Information Sheets on Shell Hydraulic Fluids*, 1996, Shell UK Ltd. Manchester, U.K.

7. *Product Data Sheets and Information Sheets on Hydraulic Fluids*, Houghton Vaughan plc., Birmingham, U.K.

8. Wilson, W., "Water-Based Hydraulic Systems for Metals Industry," *Industrial Lubrication and Tribology,* Jan.–Feb. 1992, pp. 21–23.

9. Young, K.J., and Kennedy, A., "Development of an Advanced Oil-in-Water Emulsion Hydraulic Fluid, and its Application as an Alternative to Mineral Hydraulic Oil in a High Fire Risk Environment," *Lubrication Engineering,* **49, 11,** Nov. 1993, pp. 873–879.

10. Mezger, R.H., "Design for High Water-Base Fluid," *Design News,* Jan. 21, 1980, pp. 52–56.

11. Ralph LeMar, "Planning for High Water Content Fluid," *Lubrication Engineering,* **46, 11,** Nov. 1990, pp. 737–739.

12. "High Water Content Systems for Profit Making Designs," *Hydraulics and Pneumatics,* **35, 4,** April 1982, pp. HP1-32.

13. ANSI (NFPA) T2.13.5-1991, *Hydraulic Fluid Power–Industrial Systems–Practice for the Use of High Water Content Fluids,* American National Standards Institute, New York, NY, U.S.

14. *High Water Content Fluid, Laboratory Report,* Bulletin 90009, 1984, The Oil Gear Company, Milwaukee, WI, U.S.

15. Merzger, R.H., "Hardware and Water-Based Fluids: Are They Compatible?" *Machine Design,* **51, 17,** 1979, pp. 120–123.

16. Knight, G.C., "Water Hydraulics," *Tribology International,* April 1977, pp. 105–107.

17. Protheroe, A.R., "The Case for High Water Hydraulics," *American Machinist,* Jan.1980, pp. 128–129.

18. *Hydraulic Systems Using High Water-Based Fluids,* Bulletin HWBF 2, 1999, Oilgear Towler Limited, Leeds, U.K.

19. *Dipslide Leaflet,* de la Pena Lubricants, Pershore, U.K.

20. Katzenbach, A., and Jost, R., "HFA Hydraulics in Production Machinery [in German]," *Oel Hydraulik und Pneumatik,* **35, 3,** March 1991, pp. 198–201.

CHAPTER 5

1. *Fire Safe Fluid Guide,* Technical Bulletin, Oiltech Australia Pvt.Ltd, Cheltenham, Australia, March 2000.

2. Phillips, W.D., "A Comparison of Fire-Resistant Hydraulic Fluids for Hazardous Industrial Environments, Part-1: Fire Resistance and Lubrication Properties," *Journal of Synthetic Lubrication,* **14, 3,** 1998, pp. 211–235.

3. *Fire-Resistant Hydraulic Fluids: Data Sheets,* BFPA Report No. P13-1996, The British Fluid Power Association, Chipping Norton, U.K.

4. Schmiede, J.T.; Simandiri, S.; and Clark, A.J.; "The Development and Applications of an Invert Emulsion Fire-Resistant Hydraulic Fluid," *Lubrication Engineering,* **41, 8,** April 1985, pp. 463–469.

5. Borowski, J.L., "The Use of Invert Emulsion Hydraulic Fluid in a Steel Slab Caster," *Lubrication Engineering,* **36, 11,** Nov. 1980, pp. 652–656.

6. Young, K.J., "Development and Application of Advanced Emulsion Hydraulic Fluids," SAE Technical Paper No. 951196, *SAE Transactions, Section 2,* 1996, pp. 19–26.

7. Law, D.A., "The Development and Testing of an Advanced Water-in-Emulsion for Underground Mine Service," *Lubrication Engineering*, **37, 2,** Feb. 1981, pp. 82–90.
8. Garti, N.; Feldenkriez, R.; Aserin, A.; Ezrahi, S.; and Shapira, D.; "Hydraulic Fluids Based on Water-in-Oil Microemulsions," *Lubrication Engineering*, **49, 5,** May 1993, pp. 401–411.
9. Myers, M.B., "Testing and Industrial Experience with Water-in-Oil Fire Resistant Hydraulic Fluids," *Proceedings of the International Symposium on Performance Testing of Hydraulic Fluids,* Institute of Petroleum, London, 1978, pp. 321–336.
10. *Technical Product Information TPI-459-07-96, Fire Resistant Water Glycol Fluids,* 1996, Denison Hydraulics France S.A., Vierzon Cedex, France.
11. Blanpain, G., "The Use of Polyglycols in French Coal Mines," *Proceedings of the International Symposium on Performance Testing of Hydraulic Fluids,* Institute of Petroleum, London, 1978, pp. 389–403.
12. Vardy, E.N., "The Use of Present Day Hydraulic Fluids in the Steel Industry," *ibid,* pp. 521–531.
13. Ross, G.E., "Fire Resistant Hydraulic Fluids in High-Pressure Die Casting Machines," *ibid,* pp. 533–542.
14. Natscher, J., *Hydraulic Fluids in Metallurgical Plants and Rolling Mills,* Document No. RE 00 252/09.94, Mannesmann Rexroth GmbH, Lohr am Main, Germany.
15. Fletcher B, "Caster Hydraulics at Stelco's Lake Erie Works," *Iron and Steel Engineer,* Oct. 1984, pp. 39–43.
16. Reichel, J., "Fluid Power Engineering With Fire Resistant Hydraulic Fluids–Experience with Water-Containing Hydraulic Fluids," *Lubrication Engineering,* **50, 12,** Dec. 1994, pp. 947–952.
17. Wachter, D.A.; Bishop, R.J.; McDaniels, R.L.; and Totten, G.E.; "Water Glycol Hydraulic Fluid Performance Monitoring: Fluid Performance and Analysis Strategy," SAE Technical Paper No. 952155, *SAE Transactions, Section 2,* 1995, pp. 389–398.
18. Staley, C. and Hyde, T.G., "Modern Phosphate Ester Hydraulic Fluids," *Power International,* **32, 381,** Sep. 1986, pp. 232–233.
19. Marolewski, T.A. and Hartsuch, P.W., "Properties and Performance of Triaryl Phosphate Ester Hydraulic Fluids," *Proceedings of the 47th National Conference on Fluid Power,* National Fluid Power Association, Milwaukee, WI, U.S., Vol.1, April 1996, pp. 103–105.
20. Marolewski, T.A. and Weil E.D., "A Review of Phosphate Ester Fire Resistance Mechanisms and their Relevance to Fluid Testing," *Fire Resistance of Industrial Fluids, ASTM STP 1284,* 1995, American Society for Testing and Materials, West Conshohocken, PA, U.S.
21. Goode, M.J.; Phillips, W.D.; and Placek, D.; "Triaryl Phosphate Ester Hydraulic Fluids–A Reassessment of their Toxicity and Environmental Behavior," SAE Technical Paper No. 982004, *Hydraulic Fluids and Alternative Industrial Lubricants, SP-1384,* Society of Automotive Engineers, Warrendale, PA, U.S., 1998.
22. Phillips, W.D., "A Comparison of Fire-Resistant Hydraulic Fluids for Hazardous Industrial Environments, Part-2: Stability, Fluid Life, and Environment," *Journal of Synthetic Lubrication,* **14, 4,** 1998, pp. 307–330.

23. Stark, L.R., "Status of Fire-Resistant Turbine Lubricants," *Lubrication Engineering*, **33,10,** Oct. 1977, pp. 535–537.
24. Staley, C. and McGuigan, B., "The European Use of Phosphate Esters in Steam and Gas Turbines," *Lubrication Engineering*, **33, 10,** Oct. 1977, pp. 527–534.
25. Phillips, W.D., *Developments in the Use of Phosphate Ester Fire-Resistant Hydraulic Fluids and Lubricants for Steam and Gas Turbines*, FMC Corporation, Philadelphia, PA, U.S.
26. Muller, U.J., "Hydraulic Fluids," *Ullmann's Encyclopaedia of Industrial Chemistry*, Volume A 13, V Edition, pp. 165–176.
27. ISO 7745: 1999, *Hydraulic Fluid Power–Fire Resistant (FR) Fluids–Guidelines for Use*, International Organization for Standardization, Geneva, Switzerland.
28. ANSI/B 93.5–1979 (R1988), *Practice for the Use of Fire-Resistant Fluids in Industrial Hydraulic Fluid Power Systems*, American National Standards Institute, New York, NY, U.S.
29. ASTM D 4174-1994, "Cleaning, Flushing and Purification of Petroleum Fluid Hydraulic System," *Annual Book of ASTM Standards, Volume 05.02*, American Society for Testing and Materials, West Conshohocken, PA, U.S., 1995, pp. 730–739.
30. *Technical Information Sheets on Hydraulic Fluids*, 1996, Shell U.K. Ltd., Manchester, U.K.
31. *Product Data Sheets and Information Sheets on Hydraulic Fluids*, 1994, Houghton Vaughan plc., Birmingham, U.K.
32. *Technical Data Sheets on Fire Resistant Fluids*, 1996, Quaker Chemical Ltd. London, U.K.
33. *Guide to Alternative Fluids*, Bulletin 579, 1992, Vickers Inc., Troy, MI, U.S.
34. *Reolube HYD Fire-Resistant Fluids*, 1994, FMC Corporation (U.K.) Ltd., Manchester, U.K.
35. *Reolube Turbofluids–A Guide to their Maintenance and Use*, 1997, ibid.
36. *Reolube Turbofluid 46XC*, 1996, ibid.
37. Goodman, K. C., "Living with Fire-Resistant Hydraulic Fluids," *Coal Age*, Sep. 1966.
38. *Changeover Procedure–Phosphate Esters*, Technical Bulletin, 1998, Oiltech Australia Pty. Ltd., Cheltenham, Australia.
39. Wiggins, B.J., "System Conversions for Fire-Retardant Hydraulic Fluids," *Lubrication Engineering*, **43, 6,** June 1987, pp. 467–472.
40. Castleton, V.W., "Practical Considerations for Fire-Resistant Fluids," *Lubrication Engineering*, **54, 2,** Feb. 1998, pp. 11–17.

CHAPTER 6

1. Phillips, W.D., "Fire Resistance Tests for Fluids and Lubricants–Their Limitations and Misapplication," *Fire Resistance of Industrial Fluids, ASTM STP 1284*, Totten, G.E., and. Reichel, J. (Eds), American Society for Testing and Materials, West Conshohocken, PA, U.S., 1996, pp. 78–101.
2. ANSI/(NFPA) T2.13.8–1997, *Hydraulic Fluid Power—Fire-Resistant Fluids–Definitions, Classifications, and Testing*, American National Standards Institute, New York, N.Y., U.S.

3. Phillips, W.D., "A Comparison of Fire-Resistant Hydraulic Fluids for Hazardous Industrial Environments, Part-1: Fire Resistance and Lubrication Properties," *Journal of Synthetic Lubrication*, **14, 3, 1998,** pp. 211–235.

4. Totten, G.E., and Webster, G.M., "Review of Testing Methods for Hydraulic Fluid Flammability," SAE Technical Paper No. 932436, *SAE Transactions, Section 2*, 1993, pp. 382–393.

5. Cutler, D.P., "Fire Resistant Test Methods for Hydraulic Fluids," *Proceedings of the International Symposium on Performance Testing of Hydraulic Fluids*, Institute of Petroleum, London, 1978, pp. 303–310.

6. Eedy, A.R.M.; Knight, G.C.; Shorten, G.; Soul, D.M.; Tunnicliffe, T.B.; and Wilson, A.C.M.; "A Review of the Standardization of IP Test Methods for Fire-Resistant Hydraulic Fluids," *ibid*, pp. 373–379.

7. Myers, M.B., "Testing and Industrial Experience with Water-in-Oil Fire-Resistant Hydraulic Fluids," *ibid*, pp. 321–336.

8. Bauer, K.; Guse, W.; and Hermann, P.; "A Test for the Determination of the Ignitability of Fire Resistant Fluids in Contact with Molten Metals," *ibid*, pp. 311–320.

9. Beyer R, and Bauer K, "Determination of the Fire-Resistant Properties of Pressure Fluids in Contact with Molten Metals [in German], " *Oel Hydraulik und Pneumatik*, **32, 6,** June 1988, pp. 433–437.

10. Schmiede, J.T.; Simandiri, S.; and Clark, A.J.; "The Development and Applications of an Invert Emulsion Fire-Resistant Hydraulic Fluid," *Lubrication Engineering*, **41, 8,** April 1985, pp. 463–469.

11. B.S. 6336: 1982, *Guide to Development and Presentation of Fire Tests and Their Use in Hazard Assessment*, British Standards Institution, London, U.K.

12. Reichel J, "Standardization Activities for Testing of Fire-Resistant Fluids in Europe," *Fire Resistance of Industrial Fluids, ASTM STP 1284*, Totten, G.E., Reichel, J. (Eds), American Society for Testing and Materials, West Conshohocken, PA, U.S., 1995.

13. Marolewsky, T.A., and Weil, E.D., "A Review of Phosphate Ester Fire Resistance Mechanisms and Their Relevance to Fluid Testing," *ibid*.

14. ASTM E 659-78, "Standard Test Method for Autoignition Temperature of Liquid Chemicals," *Annual Book of ASTM Standards, Volume 14.02*, American Society for Testing Materials, West Conshohocken, PA, U.S., 1995, pp. 360–364.

15. DIN 51794: 1978, *Testing of Mineral Oil Hydrocarbons: Determination of Ignition Temperature*, Deutsches Institut fur Normung e.V., Berlin, Germany.

16. Code of Federal Regulations, Title 30, Volume 1, Part 35, *Fire-Resistant Hydraulic Fluids, Section 35.20, Autogenous-Ignition Temperature Test*, U.S. Bureau of Mines, July 1, 2001.

17. Kuchta, J.M., and Cato, R.J., "Ignition and Flammability Properties of Lubricants," SAE Technical Paper No. 680323, *SAE Transactions*, 1968, pp. 1008–1020.

18. *Factory Mutual Specification Test Standard for Flammability of Hydraulic Fluids-Class No. 6930*, Factory Mutual Research, Johnston, RI, U.S.

19. AMS 3150C, *Flammability, Manifold Test*, Society of Automotive Engineers, Warrendale, PA, U.S.

20. Reichel, J., Essen, Germany, *Fire Resistance of Hydraulic Fluids,* Private Communication, March 2001.

21. ISO 14935: 1998, *Petroleum and Related Products–Determination of Wick Flame Persistence of Fire-Resistant Fluids,* International Organization for Standardization, Geneva, Switzerland.

22. ASTM D 5306-92, "Standard Test Method for Linear Flame Propagation Rate of Lubricating Oils and Hydraulic Fluids", *Annual Book of ASTM Standards, Volume 02.03,* American Society for Testing and Materials, West Conshohocken, PA, U.S., 1995, pp. 553–555.

23. 7th Luxembourg Report, Section 3.2.2, *Determination of Flame Propagation in a Fluid-Coal Dust Mixture.*

24. IS: 7895-1975, *Tests for Fire-Resistant Characteristics of Hydraulic Fluid Used in Mining Machinery,* Bureau of Indian Standards, New Delhi, India.

25. Code of Federal Regulations, Title 30, Volume 1, Part 35, *Fire-Resistant Hydraulic Fluids, Section 35.22, Test to Determine the Effect of Evaporation on Flammability,* U.S. Bureau of Mines, July 1, 2001.

26. Code of Federal Regulations, Title 30, Volume 1, Part 35, *Fire Resistant Hydraulic Fluids, Section 35.21, Temperature-Pressure Spray-Ignition Test,* U.S. Bureau of Mines, July 1, 2001.

27. 7th Luxembourg Report, Section 3.1.2 and 3.1.3, *Spray Ignition Test.*

28. ISO 15029-1: 1999, *Petroleum and Related Products–Determination of Spray Ignition Characteristics of Fire-Resistant Fluids–Part 1: Spray Flame Persistence—Hollow-Cone Nozzle Method,* International Organization for Standardization, Geneva, Switzerland.

29. Khan, M.M., and Brandao, A.V., "Method of Testing the Spray Flammability of Hydraulic Fluids," SAE Technical Paper No. 921737, *SAE Transactions, Section 2,* 1992, pp. 600–605.

30. Khan, M.M., "Spray Flammability of Hydraulic Fluids," *Fire Resistance of Industrial Fluids, ASTM STP 1284,* Totten, G.E., and Reichel, J. (Eds), American Society for Testing and Materials, West Conshohocken, PA, U.S., 1996.

31. Khan, M.M., and Brandaom A.V., "Fire Properties of Hydraulic Fluids," *Proceedings of the 47th National Conference on Fluid Power,* National Fluid Power Association, Milwaukee, WI, U.S., 1996, pp. 69–74.

32. Brandao, A.V., *Spray Flammability Parameter Calculation,* 1998, Factory Mutual Research, Johnston, RI, U.S., Private Communication, July 2001.

33. ISO 4404: 1998, *Petroleum and Related Products–Determination of Corrosion Resistance of Water Containing Fire-Resistant Fluids for Hydraulic Systems,* International Organization for Standardization, Geneva, Switzerland.

34. ASTM D 5534-94, "Standard Test for Vapor-Phase Rust Preventing Characteristic of Hydraulic Fluids," *Annual Book of ASTM Standards, Volume 05.03,* American Society for Testing and Materials, West Conshohocken, PA, U.S., 1995, pp. 726–730.

35. ASTM D 3709-89, "Standard Test Method for Stability of Water-in-Oil Emulsions Under Low to Ambient Temperature Cycling Conditions," *Annual book of ASTM Standards, Volume 05.02,* American society for Testing and Materials, West Conshohocken, PA, U.S., 1995, pp. 535–536.

36. ASTM D 3707-89, "Standard Test Method for Storage Stability of Water-in-Oil Emulsions by the Oven Test Method," *ibid,* pp. 530–533.

37. *Denison Hydraulic Fluid Standard HF –3*, Denison Hydraulics France S.A., Vierzon Cedex, France.
38. IP 290/1984, *Determination of Stability of Water-in Oil Emulsions at Ambient Temperature*, Institute of Petroleum, London, U.K.

CHAPTER 7

1. *Guidelines to the Use of Ecologically Acceptable Hydraulic Fluids in Hydraulic Fluid Power Systems*, BFPA Report No. 57-1993, The British Fluid Power Association, Chipping Norton, U.K.
2. *VDMA 24568 and 24569–Rapidly Biologically Degradable Hydraulic Fluids–Minimum Technical Requirements and Conversion From Fluids Based on Mineral oils*, BFPA Report No. P65-1995, ibid.
3. Droy, F.B., and Randles, S.J., "Environmental Impact,"*Synthetic Lubricants and High Performance Functional Fluids*, Rudnick, L.R., and Shubkin, R.L. (Eds.), Second Edition, Marcel and Decker, New York, U.S., 1999, pp. 793–805.
4. Tinger, J., "Hydraulic Oils With Accelerated Biological Degradability: For Every Application, the Right Product [in German]," *Oel Hydraulik und Pneumatik*, **38, 6,** June 1994, pp. 342–345.
5. *Information report–Recommendations for Conservation, Maintenance, and Disposal of Hydraulic Fluids*, ANSI/(NFPA) T2.13.4–1994, American National Standards Institute, New York, U.S.
6. Omeis, J.; Bock, W.; and Harperscheid, M.; "The Development of a New Generation of High Performance Biofluids," SAE Technical Paper No. 981491, *Earth Moving Industry Conference and Exposition*, Society of Automotive Engineers, Warrendale, PA, U.S., April 1998.
7. *Fuchs Technical Information Sheets*, 1997, Fuchs Petrolub AG, Mannheim, Germany.
8. "Biodegradable Lubricants," *Servo Information Update,* **3,** March 1997, Indian Oil Corporation, Mumbai, India.
9. *BFPA Survey on Ecologically Acceptable Hydraulic Fluids*, BFPA Report No. P66-1995, The British Fluid Power Association, Chipping Norton, U.K.
10. Baggott, J., "Lubricants, Biodegradable", *Encyclopaedia on Energy, Technology and Environment*, Bisco, A., and Boots, S., (Eds.), Volume 3, 1995, John Wiley & Sons, Inc., U.S., pp. 1,933–1,941.
11. ASTM D 6006-1997, "Standard Guide for Assessing Biodegradability of Hydraulic Fluids," *Annual Book of ASTM Standards, Volume 05.03*, American Society for Testing and Materials, West Conshohocken, PA, U.S., 1997, pp. 1,309–1,313.
12. Hooper, L.D., and Hoel, I.D., "Lubricants, the Environment and ASTM D 02," SAE Paper No. 961727, *SAE Transactions, Section 4*, 1996, pp. 1149–1157.
13. ASTM D 5864-1995, "Standard Test Method for Determining Aerobic Aquatic Biodegradation of Lubricants or Their Components," *Annual Book of ASTM Standards, Volume 05.03*, American Society for Testing and Materials, West Conshohocken, PA, U.S., 1997, pp. 1,149–1,155.

14. Busch, C., "Study on Rapeseed Oil as Pressure Transmission Medium [in German]," *Oel Hydraulik und Pneumatik,* **35, 6,** June 1991, pp. 506–519.
15. Bondioli, P., "Basics of Oleochemistry for Lubricant Production," *CTVO-Net First Workshop on Lubricants and Hydraulic fluids,* Eibar, Spain, 1999.
16. Fessenbecker, A., and Roehrs, I., *Additives for Environmentally More Friendly Lubricants,* Rein Chemie Rheinau GmbH, Mannheim, Germany, July 1995.
17. Kroff, J., and Fessenbecker, A., "Additives for Biodegradable Lubricants," *Proceedings* of the 59th *Annual Meeting of NLGI,* U.S. Oct. 1992.
18. Rohers, I., and Rossrucker, T., "Performance and Ecology–Two Aspects for Modern Greases," *Proceedings of the 61st Annual Meeting of NLGI,* U.S. Oct. 1994.
19. Fessenbecker, A., "Additivation of Environmentally Acceptable Lubricants," *CTVO-Net First Workshop on Lubricants and Hydraulic Fluids,* Eibar, Spain, 1999.
20. Bush, C., and Blacke, W., "Development and Investigation in Biodegradable Fluids," *SAE Technical Paper No. 932450,* Society of Automotive Engineers, Warrendale, PA, U.S., 1993.
21. Fessenbecker, A., "New Additive for the Hydrolytic Stabilization of Ester Lubricants," *Proceedings of the 10th International Colloquium on Tribology,* Technishe Akademie Esslingen, Germany, Jan. 1996.
22. Kovari, K., "Technology of Vegetable Oil Production," *CTVO-Net First Workshop on Lubricants and Hydraulic fluids,* Eibar, Spain, 1999.
23. Honary, L., "Performance of Selected Vegetable Oils in ASTM Hydraulic Tests," SAE Technical Paper No. 952075, *SAE Transactions, Section 4,* 1995, pp. 840–844.
24. *Soy-Based lubricants,* Omnitech International Ltd., Midland, MI, U.S.
25. Glancey, J.L.; Knowlton, S.; and Benson, E.R.; "Development of a High Oleic Soybean Oil-Based Hydraulic Fluid," *Feedstocks,* **4, 2,** March 1999, United Soybean Board, U.S.
26. Glancy, J.L.; Knowlton, S.; and Benson, E.R.; "Development of High Oleic Soybean Oil and Hydraulic Fluid," *SAE Technical Paper No. 981999,* Society of Automotive Engineers, Warrendale, PA, U.S., 1998.
27. Padavich, A.R., and Honary, L., "A Market Research and Analysis Report on Vegetable-Based Industrial Lubricants," SAE Technical Paper No. 952077, *SAE Transactions, Section 4,* 1995, pp. 845–854.
28. Adams, R.; Kromdyk, J.P.; and Noblit, T.; "Canola Oil-Based Fluid is Gentle on Environment," *Hydraulics and Pneumatics,* April 1999, pp. 68, 70,72.
29. Honary, L.A.T., "Potential Utilization of Soybean Oil as Industrial Hydraulic Oil," *SAE Technical Paper No. 941760,* Society of Automotive Engineers, Warrendale, PA, U.S., 1994.
30. Rhee, I., "Evaluation of Environmentally Acceptable Hydraulic Fluids," *NLGI Spokesman,* Aug. 1996, pp. 28–35.
31. "Product Review: Biodegradable Fluids and Lubricants," *Industrial Lubrication and Tribology,* **48, 2,** March/April 1996, pp. 17–26.
32. *Environmentally Acceptable Hydraulic Fluids, HETG, HEPG, HEES for Axial Piston Pumps,* Document No. RE 90221/05.93, 1993, Mannesmann Rexroth Hydromatik GmbH., Elchingen, Germany.
33. Denison Standard A-SK-30320, 1997, *Environmentally Acceptable Hydraulic Fluids,* Denison Hydraulics France S.A., Vierzon Cedex, France.

34. *112 B Biodegradable Hydraulic Fluid ISO 46 & 68,* Technical Bulletin, Schaeffer Manufacturing Company, St.Louis, MO, U.S.

35. Kessler, M., and Feldmann, D.G., "Evaluation of Application Related Properties of Hydraulic Fluids by Laboratory Tests and Experiences with Biodegradable Fluids and Field," SAE Technical Paper No. 982002, *Hydraulic Fluids and Alternative Industrial Lubricants, SP-1384,* Society of Automotive Engineers, Warrendale, PA, U.S., 1998.

36. Natscher, J., "Steel Hydraulic Engineering–Environmental Friendly New Media-New Knowledge [in German]," *Document No. RD 00 494-02/10.97, 1997,* Mannesmann Rexroth GmbH, Lohr am Main, Germany.

37. *Lubricants Data Sheets on Biodegradable Hydraulic Fluids,* 1996, Shell UK Ltd., Manchester, U.K.

38. *Environmentally Compatible Fluids for Hydraulic Components,* Document No. RE 03 145/05.91, 1991, Mannesmann Rexroth GmbH, Lohr am Main, Germany.

39. Aengeneyndt, K.D., "New Ecologically Compatible Hydraulic Pressure Fluids [in German]," *Oel und Hydraulik and Pneumatik,* **30,7,** July 1986, pp. 529–531.

40. *Ecologically Acceptable Hydraulic Fluids: Data Sheets,* BFPA Report No. P67-1996, The British Fluid Power Association, Chipping Norton, U.K.

41. Foelster, N.C., and Harms, H., "Monitoring System for On-Line Observation of Oil Properties and Evaluation of the Condition of Biodegradable Fluids," *Proceedings of the 48th National Conference on Fluid Power,* National Fluid Power Association, Milwaukee, WI, U.S., April, 2000.

42. *Aerobic Aquatic Biodegradation Test,* EPA Publication 560/6-82-003, No. CG-2000, Aug. 1982, Environmental Protection Agency, U.S.

43. Gledhill, W.E., "Screening Test for Assessment of Ultimate Biodegradability: Linear Alkylbenzene Sulfonates," *Applied Microbiology,* **30, 6,** American Society for Microbiology, Dec. 1975, pp. 922–929.

44. CEC L–33-A-93, *Biodegradability of Two-Stroke Cycle Outboard Engine Oils in Water,* Coordinating European Council, Earl Shilton, U.K.

45. *OECD 301B CO_2 Evolution Test (Modified Sturm Test),* OECD Guidelines for Testing of Chemicals, Ready Biodegradability, July 1992.

46. *OECD 203, Fish, Acute Toxicity Test,* OECD Guidelines for Testing of Chemicals, Ready Biodegradability, July 1992.

47. Chapter 8, "Environmentally Acceptable Lubricants," *Engineering and Design, Lubricants and Hydraulic Fluids Manual No. EM 1110-2-1424,* Feb. 28, 1999, Department of the Army. U.S. Army Corps of Engineers, U.S.

48. "Engineering and Design: Environmentally Acceptable Lubricating Oils, Greases and Hydraulic Fluids,"*Technical Letter No. 1110-2-352,* April 30, 1997, Department of the Army, U.S. Army Corps of Engineers, U.S.

49. Beitelman, A.D., "Environmentally Friendly Lubricants," *The REMR Bulletin,* **13, 2,** May 1996, Repair, Evaluation, Maintenance and Rehabilitation (REMR), Department of the Army, U.S.

50. Rhee, I., "Development of Biodegradable Hydraulic Fluids for Military Applications," *Third Annual Joint Service Pollution Prevention Conference and Exhibition,* Aug. 1998, NDIA, Defense Technical Information Center (DTIC), U.S.

51. Honary, L., "A Status Report on Federal and State Initiatives to Promote the Use of Biobased Hydraulic Fluids in the U.S.," SAE Technical Paper No.1999-01-2867, *SAE Transactions, Section 2,* 1999, pp. 438–442.

52. Kutter, B., and Feldmann, D.G., "Improvement of the Operating Characteristics of Marine Low-Pressure Hydraulic Machines," *Proceedings of the First FPNI-PhD Symposium 2000*, Hamburg, Germany, pp. 383–395.
53. "Roundtable Discussions: Environment Friendly Pressure Fluids [in German]," *Oel Hydraulik und Pneumatik*, **35, 2**, Feb. 1991, pp. 84–99.
54. Murr, T., "Biologically Degradable Environment Friendly Pressure Media [in German]," *Oel Hydraulik und Pneumatik*, **35, 2**, Feb. 1991, pp. 100–101.
55. Backe, W., "Aachen Fluid Power Colloquium (AFK): A Survey of the Present State and Trends in Fluid Technology [in German]," *Oel Hydraulik und Pneumatik*, **36, 1**, Jan. 1992, pp. 16–19.
56. Goode, M.J.; Phillips, W.D.; and Placek, D.; "Triaryl Phosphate Ester Hydraulic Fluids–A Reassessment of Their Toxicity and Environmental Behavior," SAE Technical Paper No. 982004, *Hydraulic Fluids and Alternative Industrial Lubricants, SP-1384*, Society of Automotive Engineers, Warrendale, PA, U.S., 1998.
57. Kiovsky, T.E.; Murr, T.; and Voeltz, M.; "Biodegradable Hydraulic Fluids and Related Lubricants," SAE Technical Paper No. 942287, *SAE Transactions, Section 4*, 1994, pp. 2,013–2,020.
58. Eguchi, R.; Ohtakem Y.; Ohkawa, S.; Iwamura, M.; and Konishi, A.; "Compatibility of Hydraulic Seal Elastomers With Biodegradable Oils," SAE Technical Paper No. 960210, *SAE Transactions, Section 5*, 1996, pp. 219–224.
59. Remmelmann, A., "Testing Method for Biodegradable Hydraulic Pressure Media Based on Natural and Synthetic Esters," *Tribology of Hydraulic Pump Testing, ASTM STP 1310*, Totten, G.E.; Kling, G.H.; and Smolenski, D.M. (Eds), American Society for Testing and Materials, West Conshohocken, PA, U.S., 1995.
60. Cheng, V.M.; Galiano-Roth, A.S.; Marougy, T.; and Berezinski, J.; "Vegetable-Based Hydraulic Oil Performance in Piston Pumps," SAE Technical Paper No. 941079, *SAE Transactions, Section 4*, 1994, pp. 734–742.
61. Ohkawa, S.; Konish, A.; Hatano, H.; Ishihama, K.; Tanaka, K.; and Iwamura, M.; "Oxidation and Corrosion Characteristics of Vegetable-Base Oil," SAE Technical Paper No. 951038, *SAE Transactions, Section 4*, 1995, pp. 737–745.
62. Totten, G.E.; Cerf, J.; Bishop, R.J., Jr.; and Webster, G.M.; "Recent Results of Biodegradability and Toxicology Studies of Water Glycol Hydraulic Fluids," SAE Technical Paper No. 972744, *SAE Transactions, Section 2*, 1997, pp. 241–246.
63. Igartua, A., et al., "Behavior of Vegetable Oil: Biodegradability, Toxicity, Recycling of Vegetable Oils, Lifecycle," *CTVO-Net First Workshop on Lubricants and Hydraulic Fluids*, Eibar, Spain, 1999.
64. Dowson, D., *The History of Tribology*, Longman Group Limited, London, 1979.

CHAPTER 8

1. ISO 4021:1992, *Hydraulic Fluid Power–Particulate Contamination Analysis–Extraction of Fluid Samples from Lines of an Operating System*, International Organization for Standardization, Geneva, Switzerland.
2. ISO 3722:1976, *Hydraulic Fluid Power–Fluid Sample Containers–Qualifying and Controlling Cleaning Methods*, ibid.

3. ISO 4406:1999, *Hydraulic Fluid Power—Fluids-Method for Coding the Level of Contamination by Solid Particles,* ibid.

4. ANSI B93.19, *Method for Extracting Fluid Samples from the Lines of an Operating Hydraulic Fluid Power System (for particulate analysis),* American National Standards Institute, New York. U.S.

5. Fitch, J.C., and Alfred, J.B., "Hydraulic Fluid Analysis: Avoiding the Pitfalls, Part 1," *Hydraulics and Pneumatics,* **40, 12,** Dec. 1987, pp. 69–71.

6. Fitch, J.C., and Alfred, J.B., "Hydraulic Fluid Analysis: Avoiding the Pitfalls, Part 2: Interpretation of Test Results Plus Do-it-Yourself On-Site Tests," *Hydraulics and Pneumatics* **41,1,** Jan. 1988, pp. 98–103.

7. Poley, J., "Oil Analysis for Monitoring Hydraulic Oil Systems, a Step-stage Approach," *Lubrication Engineering,* **46, 1,** Jan. 1996, pp. 41–47.

8. Eriction, R.W., and Taylor, W.V., "Rapid Oil Analysis," *Lubrication,* **40, 1,** Caltex Petroleum Corporation, Irving, TX, U.S., 1985, pp. 1–12.

9. Young, C.H., "Used Hydraulic Oil Analysis," *Lubrication,* **63, 4,** Texaco Inc., White Plains, NY, U.S., 1977, pp. 37–48.

10. Foszcz, J.L., "Maintaining Hydraulic Fluids," *Plant Engineering,* Dec. 2000, pp. 38–42.

11. Sasiki, A.; Sasoka, M.; Kanto Seiki, K.K.; Tobisu, T.; Uchiyama, S.; and Sakai, T.; "The Use of Electrostatic Liquid Cleaning for Contamination Control of Hydraulic Oil," *Lubrication Engineering,* **44, 3,** March 1988, pp. 251–256.

12. Sasaki, A., and Yamamoto, T., "A Review of Studies of Hydraulic Lock," *Lubrication Engineering,* **49, 8,** Aug. 1993, pp. 585–593.

13. Inoue, R., "Contamination Effects: What Happens to Proportional Valves," *Hydraulics and Pneumatics,* **37, 11,** Nov. 1984, pp. 156–159.

14. Higbee T, "Optimum Filtration for Fluid Power Systems," *Plant Engineering,* Dec. 01, 1999.

15. *Vickers Guide to Systemic Contamination Control,* Catalogue 561, 1992, Vickers Inc., Troy, MI, U.S.

16. *How to Use Pall Portable Contamination Analysis Kit,* 1999, Pall Industrial Hydraulics Ltd., Port Washington, NY, U.S.

17. *Stop Leaks,* Bulletin 394, Vickers Inc., Troy, MI, U.S.

18. *Portable Test Kit for Lubricating Oils,* Stanhope-Seta Ltd., Chertsey, U.K.

19. Williams, W.T., "Field Test Kit for Evaluating Lubricants in Service," *Lubrication Engineering,* **33, 4,** April 1977, pp.191–194.

20. Kato, T., and Kawamura, M., "Oil Maintenance Tester: A New Device to Detect the Degradation," *Lubrication Engineering,* **42, 11,** Nov.1986, pp. 694–699.

21. Geary, P.A., "Evaluation of In–Service Industrial Lubricants Through Oil Analysis Kit Methods," *Lubrication Engineering,* **40, 6,** June 1984, pp. 352–355.

22. ASTM D 4174-1994, "Cleaning, Flushing and Purification of Petroleum Fluid Hydraulic System," *Annual Book of ASTM Standards, Volume 05.02,* American Society for Testing and Materials, West Conshohocken, PA, U.S., 1995, pp. 730–739.

23. Duncan, J.P., "Planning to Use F.R "Inverts" in Continuous Miners? If so, Here Are a Few do's and don'ts to Remember," *Coal Age,* April 1973.

24. Borowski, J.L., "The Use of Invert Emulsion Hydraulic Fluid in a Steel Slab Caster," *Lubrication Engineering,* **36, 11,** Nov.1980, pp. 652–656.

25. Shade, W.N., "Field Experience With Degraded Synthetic Phosphate Ester Lubricants," *Lubrication Engineering*, **43, 3,** March 1987, pp. 176–182.
26. Phillips W.D., "Triaryl Phosphates–the Next Generation of Lubricants for Steam and Gas Turbines," *ASME Joint International Power Generation Conference, Paper No. 94-JPGC-PWR-64,* American Society of Mechanical Engineers, New York, U.S., 1994.
27. Phillips W.D., "The Conditioning of Phosphate Ester Fluids in Turbine Applications," *Lubrication Engineering*, **39, 12,** Dec. 1983, pp. 766–780.
28. *Reolube Turbofluids: A Guide to Their Maintenance and Use,* 1997, FMC Corporation (U.K) Ltd., Manchester, U.K.
29. Phillips, W.D., "The Electrochemical Erosion of Servo Valves by Phosphate Ester Fire-Resistant Hydraulic Fluids," *Lubrication Engineering*, **44, 9,** Sep. 1988, pp. 758–767.
30. Marolewski, T.A., and Hartsuch, P.W., "Properties and Performance of Triaryl Phosphate Ester Hydraulic Fluids," *Proceedings of the 47th National Conference on Fluid Power,* National Fluid Power Association, Milwaukee, WI, U.S., 1996.
31. Phillips, W.D., and Sutton, D.I., "Improved Maintenance and Life Extension of Phosphate Esters Using Ion Exchange Treatment," *10th International Colloquium Tribology,* Esslingen, Germany, Jan.1996.
32. *Texaco Fluid Management Program,* Texaco Lubricant Company, U.S.
33. *Fluid Management,* Shell Lubrication Services, Shell Lubricants, U.S.
34. *Fluid Analysis Service,* Vickers Inc., Rochester Hills, MI, U.S.

CHAPTER 9

1. *Draft Technical Guidelines on Used oil Rerefining or Other Reuses of Previously Used Oil (R9),* Basel Convention, United Nations Environmental Program (UNEP) Document.
2. *European Economic Community Council Directive 75/439/EEC of June 16, 1975 On the Disposal of Waste Oils,* Official Journal of the European Commission, **No. L 194,** July 25, 1975, pp. 23–25.
3. *European Economic Community Council Directive 87/101/EEC of December 22, 1986 – Amendment to Directive 75/439/EEC of June 16, 1975 On the Disposal of Waste Oils,* Official Journal of the European Commission, **No. L 042,** Feb. 12, 1987, pp. 43–47.
4. *Legislative Decree n° 691 of August 23, 1982, n° 95 of January 27, 1992, and n° 475 of November 9,* 1988, Regulations Concerning the Elimination of Waste Oils, Laws, Decrees and Presidential Orders, Official Journal of the Republic of Italy.
5. *Title 19, Environmental Quality Code, Chapter 6-Hazardous Substances, Part 7, Used Oil Management Act,* Utah, U.S.
6. New York Codes, Rules and Regulation, *6NYCRR Part 360-Solid Waste Management Facilities, Subpart 360-14, Used Oil,* The Division of Solid and Hazardous Materials, New York Department of Environmental Conservation, U.S., Nov. 1996.

7. Garthe, J.W., and Kowal, P.D., *Recycling Used Oils*, Department of Environmental Resources, Pennsylvania Used Oil Recycling Information Center, Harrisburg, PA, U.S.

8. Blundell, G., *Background Paper to Policy Forum: Used Motor Oil–Defining a Management Strategy for Ontario*, Provincial and State Policies on Used Motor Oil Management for the Recycling Council of Ontario, Canada, May 1998.

9. *Document on Environmental Code of Practice for the Management of Used Lubricating Oil*, Australian Institute of Petroleum and Oil Recyclers Association of Australia, page 43 of 52.

10. ASTM D 4175-1992, "Standard Terminology Relating to Petroleum, Petroleum Products and Lubricants," *Annual Book of ASTM Standards, Volume 05.02*, American Society for Testing and Materials, West Conshohocken, PA, U.S. 1995, pp. 741–742.

11. *Pennsylvania Used Oil Recycling Program–A Manual for Community Used Oil Recovery Programs*, Pennsylvania Used Oil Recycling Information Center, Harrisburg, PA, U.S.

12. Igartua A et al, "Behavior of Vegetable Oil: Biodegradability, Toxicity, Recycling of Vegetable Oils, Lifecycle," *CTVO-Net First Workshop on Lubricants and Hydraulic Fluids*, Eibar, Spain, 1999.

13. *Europalub*, Lubricants Statistics 1999, Aug. 2000.

14. Watanabe S., "Recycling Used Oils," *Japanese Journal of Tribology*, **38, 5,** 1993, pp. 631–638.

15. Klamann, D., *Lubricants and Related Products: Synthesis, Properties, Application, International Standards*, Verlag Chemie GmbH, Germany, 1984.

16. *Failure of a Member State to Fulfill Obligations–Directive 87/101/EEC–Disposal of Waste Oils*, Case C-102/97, Sep.9, 1999, Judgment of the Court (Fifth Chamber), Court of Justice of the European Communities.

17. *Environmental Guidance Regulatory Bulletin*, U.S. Department of Energy Office of Environmental Guidelines, RCRA/CERCLA Division (EH-231), Oct. 31, 1993.

18. Mouche, C., "Managing Used Oil: Sorbent Technologies Promote Compliance with Used Oil Standards," *Pollution Engineering*, Oct.1, 1995

19. *North Carolina Solid and Waste Management Act, NCGS 1304-290 to 309*, Chapter 130 A, Article 9, Solid Waste Management Act, North Carolina, U.S.

20. *Used Oil Related oil Waste Management for Transporters*, Hazardous Waste Fact sheet #4.31, Minnesota Pollution Central Agency, U.S. May 1997.

21. *The Used Oil Recycling Hand Book RG 325–Guidance for Used Oil Handling*, Texas Natural Resource Conservation Commission (TNRCC), Austin, TX, U.S., June 2000.

22. *Used Oil Management Act*, Florida Statutes, Chapter 62-710, Florida Solid Waste Management Act, U.S., March 1997.

23. ANSI/(NFPA) T2.13.4–1994, *Information Report–Recommendations for Conservation, Maintenance and Disposal of Hydraulic Fluids*, American National Standards Institute, New York. NY, U.S.

24. *Ministry of Environment and Forests Notification*, The Gazette of India: Extraordinary Part II-Section 3, subsection (i), No.364, September 6, 1995, vide GSR No.620 (E).

25. *Technical Bulletin on Environment and Collect Oils,* ROSE Foundation (Recovery of Oil Saves Environment), Mowbray, South Africa.
26. *Managing Used Oil: Advice for Small Businesses,* Publication EPA53Q-F-96-004, Office of Solid Waste, Environmental Protection Agency (EPA), U.S., Nov. 1996.
27. Wilson, B., "Used Oil Reclamation Process," *Industrial Lubrication and Tribology,* **49, 4,** July–Aug. 1997, pp. 178–180.
28. Schieppati, R, and Giovanna, F.D., "Rebirth of the Rerefining Industry in Europe: Prospects for the Future," *The Fourth ICIS-LOR World Base Oils Conference,* Millennium Britannia Mayfair, London, U.K., Feb. 17–18, 2000.
29. Morawski, C., *Used Oil Recovery Programs Expand Across Canada,* Hazardous Material Management, Southam Environment Group, Ontario, Canada, June/July 2000.
30. *Lubrication Standards L9: Recycling Technology,* SMRP Virtual Library, Society for Maintenance and Reliability Professionals, Chicago, IL, U.S.
31. *General Motors Leads the Way in Recycling Oil,* GM Press Release, March 11, 1999.
32. Blatz, F.J., "Rerefined Locomotive Engine Oils and Resource Conservation," *Lubrication Engineering,* **35, 11,** Nov. 1979, pp. 618–624.
33. Snyder, C.E., and Schenach, T.A., "Recycling of MIL-H-5606 and MIL-H-6083 Mineral Oil-Based Aerospace Hydraulic Fluids," *Lubrication Engineering,* **40, 11,** Nov. 1984, pp. 667–671.
34. Two, A.B., "Properties of Recycled Hydraulic and Lubricating Oils," *Lubrication Engineering,* **31, 2,** Feb. 1975, pp. 68–71.
35. Webb, H.R., "Establishing Oil Recovery and Reclamation Programs," *Lubrication Engineering,* **39, 10,** Oct. 1983, pp. 626–630.
36. Newman, R.T., "A Lube Oil Recycling Program on a U.S. Southern Railroad," *Lubrication Engineering,* **39, 10,** Oct. 1983, pp. 640–643.
37. Sullivan, J.R., "In-Plant Oil Reclamation–Case Study," *Lubrication Engineering,* **38, 7,** July 1982, pp. 409–411.
38. Katzenstein, W., "Industrial Oil Recycling at Chrysler," *Proceedings of a Workshop on Measurements and Standards for Recycled Oil,* NBS Special Publication 488, National Institute of Standards and Technology, Gaithersburg, MD, U.S., 1977, pp. 39–40.
39. Middleton, V.L, "Conservation and Recovery of Lubricating and Hydraulic Oils," *Iron and Steel Engineer,* Nov. 1984, pp. 54–56.
40. Siegel, R., and Skidd, C., "Case Studies Utilizing Mobile On-Site Recycling of Industrial Oils for Immediate Application," *Lubrication Engineering,* **51, 9,** Sep. 1995, pp. 767–770.
41. Bowering, R.E.; Davis, C.T.; and Brainiff, D.P.; "Managing and Recycling Hydraulic and Process Oils at a Cold Mill," *Lubrication Engineering,* **53, 7,** July 1997, pp. 12–17.
42. *Texaco Fluid Management Program,* Texaco Lubricant Company, U.S.
43. *Fluid Management,* Shell Lubrication Services, Shell Lubricants, U.S.
44. *Exxon Oil Reclamation Program,* ExxonMobil Lubricants and Petroleum Specialties Company, U.S. 1997.
45. Gressel, A., *Technologies and Methods Available for Recycling Used Oils,* Research Oil Company, U.S.

46. McKeagan, D.J., "Economics of Rerefining Used Lubricants," *Lubrication Engineering*, **48, 5**, May 1992, pp. 418–423.
47. Pyziak, T., and Brinkman, D.W., "Recycling and Rerefining Used Oils," *Lubrication Engineering*, **49, 5**, May 1993, pp. 339–345.
48. Brinkman, D.W., "Technologies for Rerefining Used Lubricating Oil," *Lubrication Engineering*, **43, 5**, May 1987, pp. 324–328.
49. Swain, J.W., "Conservation, Recycling and Disposal of Industrial Lubricating Fluids,"*Lubrication Engineering*, **39, 9**, Sep. 1983, pp. 551–554.
50. Whisman, L., "New Rerefining Technologies of Western World," *Lubrication Engineering*, **35, 5**, May 1979, pp. 249–253.
51. Dang, G.S., "Rerefining of Used Oils–A Review of Commercial Processes," *Tribotest Journal*, **3–4**, June 1997, pp. 445–457.
52. Morgan, C., "Breaking Down Barriers to Small-Scale Rerefining, While Producing Virgin Quality Base Oils Without Hydrotreating," *National Oil Recyclers Association (NORA) Annual Conference*, Nov. 1996, U.S.
53. Saunders, J., "Used Oil Refining Revolution," *Lubricants World*, Sep. 1996, pp. 20–24.
54. Kress, D., "The New Economics of Used Oil Reprocessing," *Canadian Chemical News Magazine*, June 1996.
55. Arner, R., *Used oil recycling: Closing the loop*, Solid Waste Program Manager, Northern Virginia Planning District Commission, Annandale, VA, U.S.
56. Ramsden, D.P., *Responsible Management of Used Lubricants by Closed-Loop Recycling*, Technical Paper, Evergreen Holdings Inc., Irvine, CA, U.S.
57. Schieppati, R., Viscolube Italiana SpA, Pieve Fisiraga (MI), Italy, *Private Communication*, March 8, 2001.
58. Watanabe, S., Tokyo, Japan, *Private Communication*, Feb.19, 2001.
59. Gressel, A., NORA Senior Technical Consultant, Ohio, U.S., *Private Communication*, March 2, 2001.
60. *The Used Oil Collection Act*, API Model Code, Aug.15, 1994, American Petroleum Institute, Washington D.C., U.S.
61. Watanabe S., "Realities of Used Lubricants Disposal and Some Proposal for the Reutilization [in Japanese]," *Plant Engineering*, Aug. 1997, pp. 30–35.
62. Mahdi, S.M., and Skold, R.O., "Ultra Filtration for the Recycling of a Model Water-Based Metalworking Fluid: Process Design Considerations," *Lubrication Engineering*, **47, 8**, Aug. 1991, pp. 686–690.
63. Dick, R.M., "Ultrafiltration for Oily Wastewater Treatment," *Lubrication Engineering*, **38, 4**, April 1982, pp. 219–222.
64. *Como Filtration System*, Como Industrial Equipment, Inc., Janesville, WI, U.S., 2000.
65. *Osmonics*, Minnetonka, MN, U.S., 2001.
66. *Separation Dynamics*, Troy, MI, U.S., 2000.
67. *Emulsion Controls Inc.*, Chula Vista, CA, U.S., 2000.

THE AUTHOR

M. Radhakrishnan received a bachelor's degree in mechanical engineering from the University of Madras, India, and a master's degree in tribology from the University of Leeds, U.K. He also has an MBA from Bangalore University, India. He had a long career as a designer of machine tools at the Central Manufacturing Technology Institute (CMTI), a premier institute for the development of machine tools and manufacturing technology in Bangalore. During his tenure at CMTI, he designed many hydraulic and CNC-controlled machine tools; notable machine tools include a diamond turning machine, an internal grinder, a large centerless grinder, a CNC vertical milling machine, and special purpose machines for automotive industries. In addition to machine tool design, tribology of machine tools was his allied professional interest, particularly hydrostatic and aerostatic bearings and cutting fluids. He was also closely associated with standardization activities relating to lubricants, as well as with many training programs in hydraulics, lubricants and lubrication, machine tool design, and bearings. He has written many articles on hydraulic fluids and lubricants for the use of practicing engineers. Currently he is a consulting engineer in tribology and machine tool design.

INDEX